GLOBAL CLASSIC LANDSCAPE DESIGN EXPLORATION HIGHLIGHTS

Global Classic Landscape Design Exploration Highlights

全球经典景观设计探索集锦 III

《景观设计》杂志社 编

大连理工大学出版社

图书在版编目（CIP）数据

全球经典景观设计探索集锦：全4册 /《景观设计》
杂志社编. -- 大连：大连理工大学出版社，2011.9
　　ISBN 978-7-5611-6520-1

　　Ⅰ.①全… Ⅱ.①景… Ⅲ.①景观设计—作品集—世
界—现代 Ⅳ.①TU-856

中国版本图书馆CIP数据核字(2011)第182901号

出版发行：大连理工大学出版社
　　　　　　（地址：大连市软件园路80号 邮编：116023）
印　　刷：利丰雅高印刷（深圳）有限公司
幅面尺寸：245mm×245mm
印　　张：60
字　　数：1300千字
出版时间：2011年9月第1版
印刷时间：2011年9月第1次印刷
策划编辑：苗慧珠
责任编辑：刘晓晶
责任校对：万莉立
版式设计：王　江　赵安康　张建实

ISBN 978-7-5611-6520-1
定　价：880.00元（全4册）

电　话：0411-84708842
传　真：0411-84701466
邮　购：0411-84708943
E-mail:dutp@dutp.cn
http://www.landscapedesign.net.cn

目录 Contents

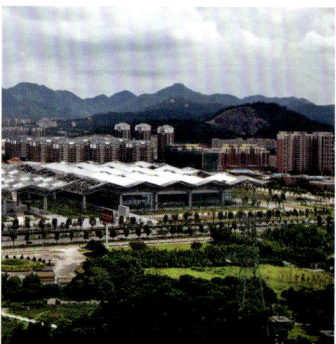

公共空间 _ Public Space

滨水空间 _ Waterfront Space

度假区/酒店 _ Resort/Hotel

目录 Contents

道路 _ Path

公共空间

工业历史的记忆 —— 钢铁博物馆

Memories of Industrial History — The Museum of Steel

撰文：Roderick Wyllie　　翻译：张璐

1986 年，蒙特雷市政府收回了占地 15 000 ㎡ 的废弃钢铁厂。多年后，Surfacedesign Inc. 与 Harari Arquitectos 联手将摇摇欲坠的高炉和一片棕地改建成一座现代历史博物馆。博物馆坐落在现代化的芬迪多拉公园中，每年接待超过 200 万的来访游客。博物馆向人们讲述着蒙特雷市辉煌悠久的钢铁生产历史，令老一辈回味，更使年轻人铭记。

该项目就地取材，将创新设计与现代建筑融为一体（格里姆肖建筑师事务所设计），为古老的遗址增添了 21 世纪的活力和气息；因地制宜，进行屋顶绿化并安装了雨洪系统；在尊重原有建筑格局的同时为景观设计提供了新的解决方案。

钢铁博物馆的景观设计充分体现了钢铁产业曾经的荣耀，突显了其在众多景观之中的重要位置。整体景观设计依托于高 70m 的高炉进行，并且不断补充新的设计元素。展现钢铁产业的发展历程是博物馆设计的主要目标，因此从原工厂回收的钢铁被广泛地用于公共广场、喷泉及露台的设计中，例如户外展区就采用了富含矿石的铁质围栏加以装饰。在挖掘过程中发现的体积庞大、外形不规则的铁质品则被用做石阶及其他景观的装饰。该设计方案既就地取材——将原有的工业遗留物进行回收利用，又通过应用环保技术恢复了当地的生态环境。

博物馆的边缘修建了一些小排水沟以帮助排放暴雨时产生的积水，这些排水沟与以往用来排放钢铁生产过程中的副产品的水道十分相似。水生植物和大型湿地植物不仅对当地的生态环境起到了保护作用，还

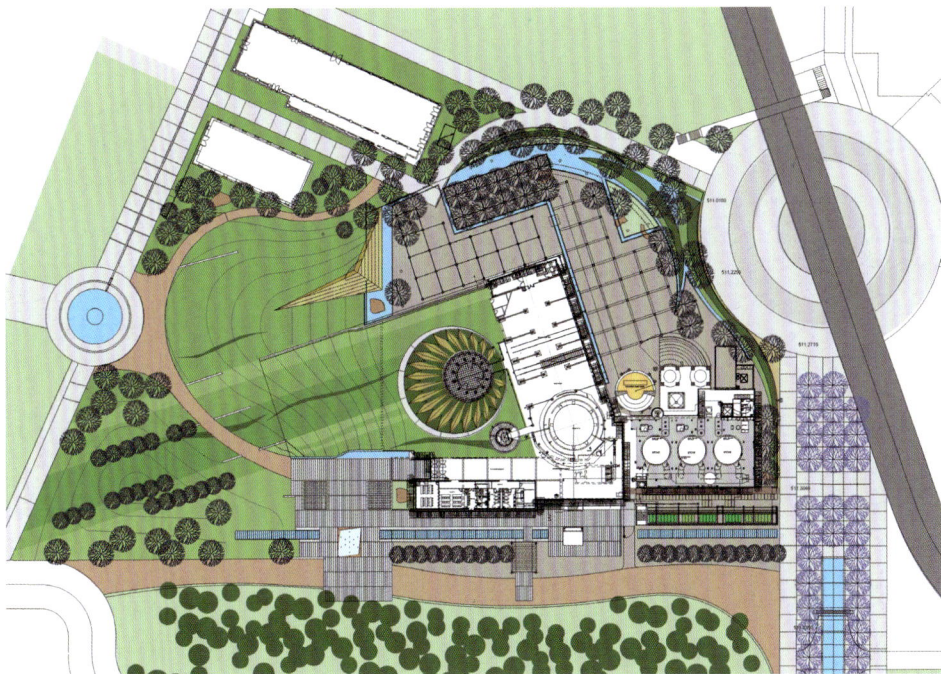

总平面图（图片提供：Surfacedesign Inc.）

1　由碎石和回收的矿石所组成的雾化喷泉景观（图片提供：Abigail Guzman Tamex）

2　水道尽头的小树林呈现出动中有静的别样景观（图片提供：Abigail Guzman Tamex）

3　回收利用原场地的金属板，突出了场地的历史特征（图片提供：Abigail Guzman Tamex）

4　绿地将人们引至主入口（图片提供：Paul Riveria）

5　钢铁博物馆夜景（图片提供：Paul Riveria）

能过滤流入蓄水池的雨水，蓄水池则为旱季的灌溉提供了保障。

　　该项目的景观设计还包括与博物馆毗邻的公共空间中的两处水景。在广场上，原来覆盖在主大厅外的铁板被改造成阶梯状的瀑布式水道。200m 长的水景向人们展示着曾经用火车装卸原材料的足迹，人们可以沿着此处水景欣赏到远处雨水花园的景致。另一处水景位于博物馆的入口处，阶梯式水道的尽头是一座雾化喷泉，石块中的矿石清晰可见。人们不禁产生错觉，

仿佛置身于提炼矿石时所产生的具有腐蚀性的热气中，而取代热气的是喷洒出的清凉水汽。微风将水汽吹向广场，给忍受蒙特雷干燥炎热气候的游人们带去清凉和喜悦。

　　博物馆大范围、高密度的屋顶绿化是拉丁美洲最大的屋顶绿化系统，缓解了新建筑带来的视觉冲击。原有的高炉矗立在新建广场的空地上。在较高的新建筑屋顶上，依照屋顶结构种植了很多耐旱景天属植物，看上去就像悬浮的钢铁圆盘。游客可以登上圆形观景

台将四周景色尽收眼底，并且还能看到远处的马德雷山，巍峨的山脉与小山似的屋顶遥相呼应。向下望去是一片自然生长的草场，不禁使人联想到这里在作为工业用地前的状态，这片草场还对退化土壤进行了生物修复。

　　该项目的景观设计将可持续性历年放在首位，通过彻底改造原有的工业高炉以及融合了与建筑景观相协调的环保技术，诠释了博物馆所在地的历史用途，并展现出其未来的艺术价值。

In 1986, the city of Monterrey, Mexico reclaimed an expansive 1.5 hectare brownfield site of a former steel production facility. Many years later, Surfacedesign Inc. collaborated with Harari Arquitectura y Paisaje to transform a decommissioned blast furnace and a brownfield site into a modern history museum. Located at the center of the modern Parque Fundidora, which receives more than two million visitors per year, the Museo Del Acero Horno3 narrates the story of steel production both to the generations who remember the history of the site and to younger visitors who may be unaware of the region's legacy.

Borrowing from materials endemic to the site, innovative landscape design weaves together with modern architecture (Grimshaw Architects) to usher an old relic into the 21st century. Environmentally sensitive technologies—such as green roofs and a storm water collection system—offer a new approach to the landscape while respecting the original context.

The landscape for the Museo Del Acero Horno3 expresses the spirit of the site's former industrial glory and celebrates its position within the surrounding dramatic landscape. The overall landscape design emphasizes the physical profile of the 70-meter furnace structure while complementing the modern design of the new structures. The history of steel is an important narrative element throughout the site, and thus steel, much of it reclaimed from the site (such as the ore-embedded steel rails used to define the outdoor exhibit spaces) is used extensively to help define public plazas and delineate fountains and landscaped terraces. Large, free-formed steel objects unearthed during excavation were incorporated as stepping stones and other features. The design approach melds industrial site reclamation—and the adaptive re-use of on-site materials—with ecological restoration through the use of green technologies.

All of the storm water runoff within the site's boundaries is treated in a series of on-site treatment runnels. These surround the exhibition areas and reinterpret the former industrial canals that once moved steel production by-products within the site. Aquatic plants and wetland macrophytes bio-remediate and treat storm water before it enters an underground cistern where it is stored for dry season irrigation.

Two water features are integral to the narrative of the project, while helping to define and locate the public space adjacent to the museum. In the main esplanade, the steel plates that formerly clad the exterior of the main hall were repurposed into a stepped canal over which water cascades. The 200-meter-long feature alludes to the tracks used daily

to train in the thousands of tons of raw materials that were off-loaded in this location, and serves as a visual connection to the rain garden in the landscape beyond. At the museum's entrance, the stepped canal culminates in the misting fountain, a grid of rocks visibly embedded with ore. This trompe l'oeil evokes the caustic heating process once used to extract ore, but instead of steam it generates a cooling mist that blows over the plaza—a pleasant surprise for visitors in Monterrey's hot and arid climate.

The use of green roofs (extensive and intensive) over the museum—which comprises the largest such roof system in Latin America—helps to reduce the visual impact of the new buildings. The existing furnace rises from this newly created ground plane. On the higher roof, a variety of drought-tolerant sedums have been arranged according to the structural roof patterns of the new architecture, and

are contained by what appears to be a floating steel disk. A circular viewing deck allows visitors to take in the expanse of surrounding regional landscape, including the distant Sierra Madres, which are echoed in the roof's mounded shape. Below, a less constrained meadow of tall grasses—an abstraction of the native landscape—creates a connection to the landscape's pre-industrial context both functioning as a bioremediation for degraded soil and increasing thermal benefits for the new structure.

Principals of sustainability are at the core of the landscape design of the Museo Del Acero Horno3. By thoughtfully repurposing found industrial artifacts and incorporating new green technologies that work in concert with the architecture and the greater landscape, the designers have created an outdoor exhibition space that interprets the area's historic uses while celebrating artistic opportunities for the future.

1　金属栏杆近景（图片提供：Abigail Guzman Tamex）

2　初看钢铁博物馆，宛如一件历史悠久的手工钢铁制品
　　（图片提供：Surfacedesign Inc.）

3　铺设主步行道的高炉耐火砖与混凝土和碎石体现了可
　　持续性（图片提供：Abigail Guzman Tamex）

4　景天属植物种植在景观台周围，仿佛高炉中燃烧的熊
　　熊火焰（图片提供：Paul Riveria）

5　露台景观（图片提供：Abigail Guzman Tamex）

绽放·新鲜 —— Corus鲜花凉亭

Unfurling & Fresh — Corus Fresh Flower

撰文：Tokin Liu　　图片提供：Keith Collie　　翻译：李沐菲

这座由夸张而绚丽的黄色花瓣组成的鲜花凉亭是专门为2008年伦敦建筑节而特别设计的可移动建筑，可以根据活动需求转移到任何不同的地点，随时创造出一个独特的公共空间。

受Corus及2008年伦敦建筑节主办方的委托，屡屡获奖的设计团队Tonkin Liu负责设计，并在Adams Kara Taylor的协助下完成了该作品。鲜花凉亭极具独创性，充分展示出在创造这样一个既易于拆卸又便于移动的建筑结构时钢材的可塑性。

鲜花凉亭艳丽喜庆的外观是对本届建筑节"新鲜"主题的极致诠释。11片逐渐展开的花瓣环绕着主茎，花瓣高度从2.8m～4.5m不等，主茎设有可进行活动的平台。这些花瓣构成一个面积为97㎡的空间，人们可以在此进行各种表演和辩论活动。游人可以从各个花瓣间的空隙进入，每个空隙的大小和高度都各不相同。白天，鲜艳的黄色花瓣在阳光的照耀下愈显亮丽；夜晚，主茎散发出的光芒使整个凉亭熠熠生辉。

Corus是一系列"新鲜思想"主题活动的发起人，他与伦敦建筑节的主办方通力协作，力求在本届建筑节上呈现出一个点睛之作。

为响应这一主题，Tonkin Liu从印度Rath Yatra寺的设计中汲取了灵感。Rath Yatra寺的一部分建于巨大的木质车轮之上，每年都会举行一次市内巡游。鲜花凉亭每到一处就预示着本届建筑节又进入到一个新的阶段，也将成为各种活动的焦点所在。这种结构不但可以作为举办各种庆典活动的场地，而且绽开的花瓣本身就是一件奇观。

这项设计充分体现出客户、设计团队、项目管理者以及施工团队的通力合作，通过他们的合作，这一需求颇高的项目才得以完美地实现。

Flaunting bright yellow voluminous petals, the fresh flower pavilion is a moveable structure designed especially for the Festival of Architecture 2008 that will transform public spaces on the route of the festival as it moves through each fresh location.

Designed by award winning architects Tonkin Liu, supported by Adams Kara Taylor and commissioned by Corus and LFA2008, the Fresh Flower pavilion is an innovative structure, which showcases the creative possibilities of steel design whilst creating an easily demountable movable structure.

Joyful in its appearance, the Fresh Flower pavilion is the ultimate expression of the 2008 festival theme FRESH. It comprises 11 unfurling petals, varying in height from 2.8 to 4.5 metres, which rise from the ground, arching around a central stalk that creates the stage. The unfurling arrangement of its petals creates a sheltered space measuring 97 sqm in which to stage public performances and debates. Visitors enter the pavilion via gaps between the petals, which vary in height and proportions. In the day the bright yellow petals will catch and glow with the sunlight and at night lights running the length of the central stalk will illuminate the pavilion.

Corus, a series sponsor of the Fresh Thinking theme worked in collaboration with the London Festival of Architecture to develop the brief to create a unique piece of signature architecture for this year's festival.

Responding to the brief, architects Tonkin Liu were inspired by the precedent of the Rath Yatra temple in India, part of which is mounted on huge wooden wheels and paraded amongst a procession across the town once a year. The arrival of the Fresh Flower at its new location will mark the transition of the festival to a new hub and provide a focal point for activities. It will become an emblem to signify the journey of the festival as the hub of activity moves from one hub to another. The structure will provide a venue for the celebration of architecture and the pavilion unfurling will be an event in itself.

The design is the result of an intense collaborative process between the client, design team, project management and assembly team set against an ambitious programme.

蒙得维的亚美国大使馆前的新公共空间
New Public Spaces for the U.S. Embassy in Montevideo

撰文：Jimena Martignoni　　图片提供：Roberto Schettini　　翻译：李沐菲

乌拉圭是南美洲最小的国家之一，其首都蒙得维的亚依世界上最宽的河流之一——柏拉特河口的海湾而建。该城市的滨水地区被称为"La Rambla"，始建于20世纪30年代，由当地的一位建筑师设计，该地区全部选用粉色花岗岩建成，是蒙得维的亚的标志性建筑之一，也是该城市最受欢迎的公共空间。

美国大使馆由建筑师贝聿铭于1970年建成，位于Rambla和几条林阴大道的前面。出于对安全方面的考虑，尤其是9·11事件后，使馆周围增设了许多白色的护栏，不仅视觉效果欠佳，而且打破了人行道与街道的原有格局，也改变了这一地区原有的交通系统。为了使这一地区更具吸引力并提供舒适的步行环境，美国大使馆、蒙得维的亚市政府以及乌拉圭建筑协会于2001年联合举办了一项竞标，希望能够设计出一种更加合适的护栏，为周围环境带来一种新鲜感，并且优化周边的公共环境氛围。建筑师Fernando Fabiano遵循此理念，以一种简洁而与周围建筑完美融合的设计赢得了该项目的竞标。

1　三个"L"形墙体给入口处增添了明显的标记，而这里曾经是一些不美观的围栏

2　新建的人行道夜景，原有的两个喷泉被重建并重置于此

3　阶梯连接了主空间与扩展空间

设计师在使馆周围设计了一系列的步行空间。其主要空间就是与使馆建筑的正面及 Rambla 平行的一条人行道，街道两侧分别点缀着棕榈树和造型细长优雅的路灯；次级空间就是侧面与其垂直相交的街道。这两个空间的路面都是由粉色花岗岩铺设而成，与原 Rambla 所用的石材出自同一地区，以保持其整体的一致性。

为了重新设置这条人行步道的轴线，设计师将原使馆右前方的人行道向 Rambla 的方向进行最大化的拓宽工作（蒙得维的亚市政府也参与了这一道路的拓宽工作），现在的人行道就位于原街道的位置。

设计师对原有的两座喷泉进行了重建，并改设在人行道两端的入口处。3 个高约 0.8 米的 "L" 形墙体成为入口处明显的标志物，这些 "L" 形墙体由粗糙的花岗岩材料制成，面向人行道而建，巧妙地围合出一些小型的植物种植区。这些新的建筑元素替代了使馆周围原有的护栏，而在使馆前面安置的花岗岩矮墙，既遮挡了原有的一些护栏又增加了新的植物种植区。

此处经过翻新改造的空间，成为了这座城市中独具魅力的场所之一。这个曾经中规中矩并带有些许古板意味的 "公共空间"，如今即便是以最完美的标准来衡量，依然可以说是一处设计成功且非常舒适的公共空间，相较于同类的建筑物，其周围的空间设计也是具有代表性的。

Uruguay is one of the smallest countries of South America and its capital city, Montevideo, is developed along a bay that opens up to the Plate River, the world's widest. The city waterfront, called "La Rambla", was built in the 1930s by a local architect; entirely made of local pink granite, this piece has become the one trademark of Montevideo as well as its most beloved public space.

The US Embassy (I. M. Pei, 1970) lies in front of the Rambla and a number of linear landscaped boulevards. However, due to security measures especially reinforced after the September 11 attacks, the whole perimeter of the building was flanked by white bollards which physically and visually interrupted sidewalks and streets, thus modifying both the pedestrian and vehicular system of the area. In order to create a more appealing and pedestrian friendly environment for the site, the US Embassy, the Montevideo's City Government and the Uruguayan Society of Architects called for a national competition in 2001 that would create more adequate barriers while providing the neighborhood with new, improved public spaces. Architect Fernando Fabiano won the competition with a quite simple layout that also incorporated an adjacent portion of land.

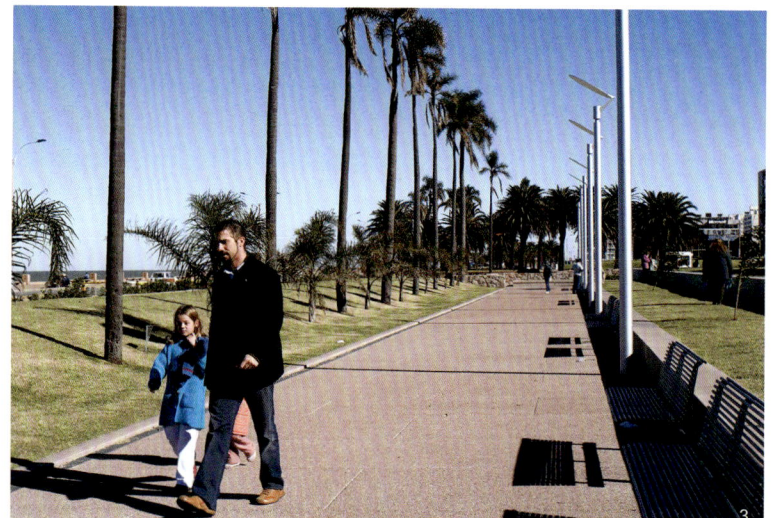

The project basically generates a sequence of pedestrian spaces developed around the embassy. A main walkway, parallel to the embassy's front façade and the Rambla, is punctuated with an existing row of palms (Arescastrum romanzoffianum) and opposite svelte light posts; a secondary space opens along the lateral perpendicular street. The pavement surfaces of both spaces were finished with the same pink granite as the historical Rambla and extracted from the original quarries.

In order to lay out this new pedestrian axis, the street that ran right in front of the embassy before was shifted the full width of the sidewalk further towards the Rambla (the City Government had anticipated this retracing as something possible) and now the path lies exactly where the street was.

Two historical fountains were restored and relocated at the beginning and end of the path and a series of three 30 inches-tall L-shaped walls demarcate the main access. Made of rough granite, these walls open towards the path and bracket small planted areas. These new elements substitute the profusion of bollards that dominated the area in the past; the rest of the bollards, which were aligned in front of the embassy, were enfolded by more granite short walls that also retain new planted areas.

This renovated place has become an inviting and quite attractive spot of the city. What used to be a conditioned and unpleasant so-called "public space" is now one that, even when still calls for complete acceptation, offers a more amiable image not very usual for this kind of controversial political space that any embassy represents nowadays.

1　人们坐在长椅上就可以欣赏到远处的河流与 La Rambla 的景色
2　大使馆前方的步行道空间，棕榈树、长椅和灯柱塑造出和缓的生活节奏
3　原有的人行道被重新设计后，成为了一条高出地平面约 0.5m 的草地坡道
4　棕榈树与灯柱为横向设计的空间增添了一些平衡感
5　人们充分利用这处空间——正在这里练习足球的孩子们
6　大使馆前方的步行空间——棕榈树、长椅和灯柱塑造出和缓的生活节奏

华盛顿国家大教堂

Washington National Cathedral

撰文 / 图片提供：Michael Vergason 景观设计有限公司　　翻译：谷晓瑞

　　1995 年，Michael Vergason 景观设计有限公司与建筑师 Torti Gallas 和一些合作伙伴最先参与到国家大教堂周边环境的总体规划中。该规划蓝图兼顾了场地附近 4 所学校的需求以及大教堂自身对游客和朝拜者的重要性。该设计以 1910 年和 1924 年奥姆斯特德兄弟的总体规划为基础，并做了细致的改良。万圣节协会从 1916 年开始负责大教堂周边地区的管理工作，工作的主要内容包括对主教庄园、香草别墅庄园、教堂西侧区域和奥姆斯特德森林的修缮和装点。此次规划工程包括神学院花园、教堂北车库以及万圣节露天剧场。设计师考虑到国家大教堂需要满足游客和朝拜者的需求，在设计上力求使教堂使命与周围具有历史底蕴的环境和谐共融。

万圣节露天剧场

　　国家大教堂坐落于美国首都华盛顿，总占地面积约 20 万平方米，其中万圣节露天剧场的占地面积约 4000 平方米。万圣节露天剧场采用现代化的设计理念，将原址的历史底蕴与大教堂中现存的建筑完美地结合在一起。万圣节露天剧场依傍于阿尔班山的奥姆斯特德森

林——现存为数不多的拥有大量橡树和山毛榉的森林，万圣节露天剧场的北面是大教堂和主教庄园。在 1924 年的总体规划中，奥姆斯特德兄弟曾构想在这里建一个露天剧场，而景观设计师恰好再次诠释了这一构想。这个露天剧场既修复了奥姆斯特德森林，也为整个大教堂增添了一道亮丽的风景线。

"森林里茂密的枝叶仿佛可以扫去城市里的尘嚣，使到访的人可以得到心灵和精神上的净化"——弗雷德里克·劳·奥姆斯特德二世（1932 年）。

奥姆斯特德二世曾认为南林（South Woods）是拜谒大教堂前做准备的一个好地方，如今景观设计师的杰作更充分地证明了这点。来大教堂朝拜的人可以在此静静地冥想，或者像在主教堂里其他的小教堂和广场中一样，祈祷、冥想和集会。

万圣节露天剧场是拯救奥姆斯特德森林的成功典范。由于该项目是基于 1924 年奥姆斯特德二世规划中的设计，这就要求万圣节露天剧场的设计既要按照客户的计划来完成，同时还要兼顾原址特殊的历史背景。这个露天剧场为集会和冥想提供了一个安静的环境，保留了高耸入云的大树，并巧妙地处理了积水问题，同时在视觉与意境上保持了与主教堂的和谐。

从 20 世纪开始，这个斜山坡就是一个非正式集会的地方，惟一保留至今的建筑是山脚的一个舞台。该场地的西面地势平缓而空旷、东面树木茂密、地势较陡。景观设计师发现舞台的中心与主教庄园的暗影殿堂

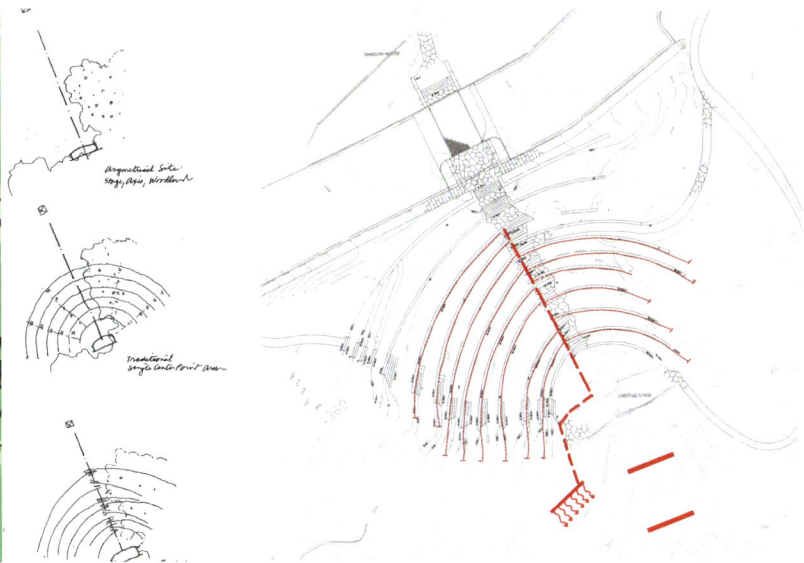

（Shadow House）和主教南塔在一条线上，形成了一种非对称的几何形状。

舞台没有因为要彻底翻新而重建，相反，设计师却将它保留了下来，这一决定使这里形成了强烈的视觉对比。正如奥姆斯特德二世蓝图中描绘的那样，景观设计师采用了传统的希腊罗马式剧院风格，在中心轴区域有所不同，两侧的阶梯随地势形成了不对称的效果。这种全新的设计仿佛位于山体裂缝之间，与暗影殿堂（Shadow House）和南塔的中心轴保持一致，这种设计创造了一种新的哥特式风格的剧院。

由弗雷德里克·劳·奥姆斯特德二世设计的奥姆斯特德森林，是阿尔班山上现存的最后一片橡树和山毛榉树林，景观设计师也为森林的修复做出了贡献。剧场依偎于森林旁，占据着重要位置。设计师针对这样一个特殊位置，利用缓坡阶地来减慢地表径流、吸收和过滤雨水，并将雨水导入到森林低处的溪谷中，对雨水控制起到了良好的作用。

考虑到奥姆斯特德森林的整体视觉效果，设计师在广场内全部种植本土植物，最大限度地发挥了本土植物的价值和美感；而从长远角度来说，这将减少广场对水的需求量。

保留舞台的决定是此次工程中基本资源合理分配的重要议题，也是其余设计定稿和参考的核心导向。设计师在选材上充分考虑了主教庄园和奥姆斯特德森林，使之与教堂周边环境形成了互补。设计师选用一块纹理细致的大卵石来迎合露天剧场多曲线、多角度的特点，同时与周边淳朴的田园气息相吻合。设计师将碎石铺在墙角处，用这样一种简单而优雅的方法来处理雨水径流；这里既是排除聚集雨水的区域，又是便捷的蓄水带。最后，再栽种上本土植物，可以减少初期的灌溉和维护需求。

国家大教堂北车库

国家大教堂北车库一方面为来访的游客提供方便，另一方面也解决了附近的威斯康星大街的交通问题。设计师在保留并美化了大教堂地区原貌的基础上设计了这个车库，减缓了教堂新通道的交通压力，并形成了一个视野开阔的环形步道。沿威斯康星大街一侧的是多孔的围墙，这使得外界可以更加清晰地欣赏到大教堂区域的景致。北车库的规划设计巧妙地融入了大教堂原有的环境之中，既保留了其原有特色，又美化了沿威斯康星大街一侧的风景。

Michael Vergason Landscape Architects, Ltd. initiated work on The National Cathedral in 1995 with the development of the Master Plan for the Close with architect Torti Gallas and Partners. The Plan provides a framework that balances the needs of the four schools of the 57 acre Close with the impact of substantial visitation and the fundamental purpose of worship associated with the Cathedral itself. The work was measured carefully against the Olmsted Brothers 1910 and 1924 Master Plans. MVLA continues to serve as the Landscape Architect for All Hallows Guild, the organization responsible for the stewardship of the Close since 1916. Work for The Guild includes renovations and additions to The Bishop's Garden, The Herb Cottage Garden, the West Front, and the Olmsted Woods. New work includes the College of Preachers garden, the North Garage and Bus Parking, and the Amphitheater. As The National Cathedral continues to grow to meet the needs of visitors and worshippers, MVLA has strived to balance these needs with the historic character of the Close.

All Hallows Amphitheater

The All Hallows Amphitheater at The National Cathedral is located in Washington, DC on one acre within the larger 54-acre Close. The All Hallows Amphitheater at The National Cathedral builds upon the landscape history and existing context of the Cathedral Close with a distinctly contemporary design solution. It is nestled next to the Olmsted Woods, one of the few remaining stands of an extensive oak and beech forest on Mount St. Alban, and immediately south of the Cathedral and Bishop's Garden. The Landscape Architect's design reinterprets the Olmsted Brothers' 1924 Master Plan, which envisioned an amphitheater on the site, it completes the Olmsted Woods Restoration, and it enriches the ensemble of landscapes of the Close.

"The great sweeping branches of the trees seem to brush off, as it were, the dust of the city, so that one at last reaches the Cathedral cleansed in mind and in spirit." —Frederick Law Olmsted, Jr., 1932.

The current Landscape Architect's work reinforced

Olmsted's view of the South Woods as an important part of the preparation for visiting the Cathedral. Contributing to the Cathedral's Pilgrimage Program, it created a place that allowed for silent contemplation and complemented the Cathedral's existing chapels and gardens used for prayer, meditation and gathering.

The All Hallows Amphitheater at The National Cathedral represented the completion of the highly successful Olmsted Woods restoration. Located on the site designated by the Olmsted Plan of 1924, the project demonstrated sensitivity to the site and history while responding to the client's program.

The Amphitheater design created a quiet backdrop for gathering and contemplation, preserved existing tree canopy, sensitively managed storm water on site, and maintained the visual and spiritual connection with the Cathedral.

The sloped hillside had been used informally as an assembly space since the early 1900s but the only permanent built structure was a stage at the lower reaches of the site. The site had gentle, open topography to the west of the centerline and tree cover and sharp topography to the east. The Landscape Architect recognized that the centerline of the stage aligned with the Shadow House of the Bishop's

Garden and the Cathedral's South Tower, creating a distinctly asymmetrical geometry.

The Landscape Architect's decision to retain the stage as opposed to rebuilding it to focus on the clearing necessitated a split vision of the site. The Landscape Architect took the traditional Greco-Roman amphitheater typology depicted in the Olmsted drawings and broke it along its center line, adapting the terraces asymmetrically to the variable conditions on either side of that spine. The new design located the primary circulation within the seam of the break, aligned with the axis of the Shadow House and South Tower.

This move created a new typology for the Amphitheater built on the form of the Gothic arch.

The Olmsted Woods, designed by Frederick Law Olmsted, Jr., is the last vestige of an extensive oak and beech forest on Mount St. Alban. The Landscape Architect's work reinforced the overall restoration plan for the Woods. The Amphitheater sits in an important location adjacent to the Woods. The Landscape Architect responded to this sensitive location by designing the Amphitheater to responsibly manage storm water on site, slowing surface flow through the terrace grading, collecting and filtering rainfall at each terrace, then transporting it and providing a slow release through spreaders in the woodland stream valley below.

Consistent with the vision for Olmsted Woods, the Landscape Architect specified all plantings within the Amphitheater be native species. This provided extended education on the value and beauty of using native plants and will reduce water demands and maintenance over the long-term.

The early decision to retain the stage was a fundamental resource allocation issue and central driver in all of the design decisions that followed. The Landscape Architect chose materials that complemented the existing Close landscape, relating to the Bishop's Garden and Olmsted

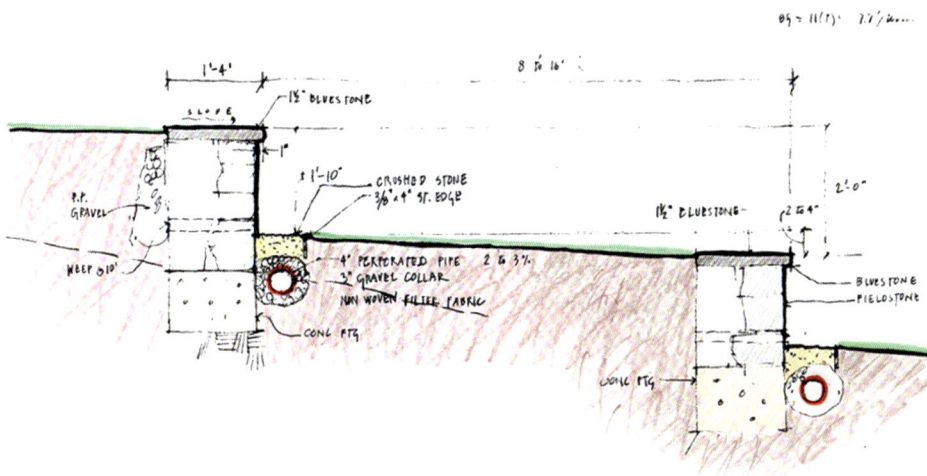

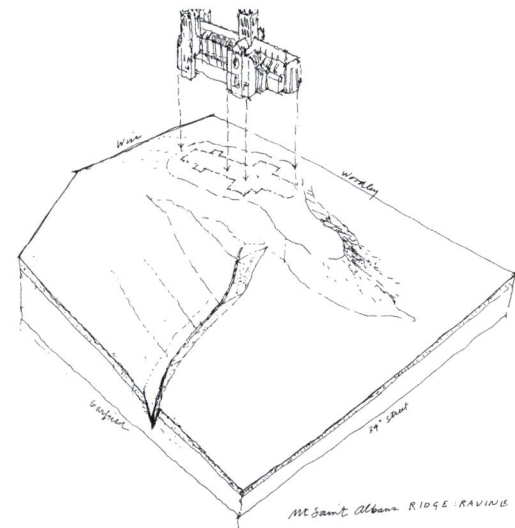

Woods. The use of a fine-textured fieldstone fit the variable curves and angles of the amphitheater and conveyed the refined rusticity found throughout the Close. The Landscape Architect also addressed storm water runoff in a simple yet elegant way, using a crushed gravel edging panel at the base of the walls. This serves as a drainage area capturing storm water and a mow strip allowing for easy maintenance. Finally, the use of a native plant palette will reduce watering and maintenance after the initial installation period.

The National Cathedral North Garage

The National Cathedral North Garage improves accessibility to the Cathedral for visitors while helping traffic management in the neighborhood and along Wisconsin Avenue. The project preserves and enhances the general appearance of the Close while integrating the new garage architecture into the landscape. The new arrival strategy at the Cathedral alleviates traffic congestion while framing views and clarifying pedestrian circulation. Increased porosity in the landscape along Wisconsin Avenue opens the Close more to the public right of way. The project incorporates the landscape more successfully into the experience of the Close, maintains its distinct character, and strengthens its presence along Wisconsin Avenue.

中山国际会展中心

Zhongshan International Exhibition Center

撰文：Elizabeth Shreeve & John Wong　　图片提供：汤姆·福克斯（SWA）　　翻译：张治程

1

中山国际会展中心犹如一座大型公园，景观别致、功能完善、布局和谐；由贸易展厅、综合展厅和会议中心等部分组成，能够满足不同的展会需求。

该项目场地的地势低洼，东西两侧茂密的森林掩映着宏伟的建筑群。森林中的树种丰富、层次分明，茂密幽深、环境宜人。该项目两大主要的景观元素——VIP会员专用的迎展广场和位于博爱路的公共广场，形成了两个明显的露天区域。露天区域与主建筑区相连接，而长鸣河和浅水渠则构成了环绕主建筑区的"护城河"水系统。

公共广场位于展区的东北部，广场上设有屋顶棚架，适合户外展览。公共广场的东部缓缓倾斜，以直线形水景为界，形成一座圆形的露天广场。

环绕主建筑区的"护城河"（即河流生态系统）虽然水位不高，却兼具美化环境与收集雨水的作用。长鸣河流经会展中心的西侧与南侧，在干旱季节，河体几近干涸，河道中的鹅卵石清晰可见；雨季时，长鸣河能够起到雨水分洪的作用，由步道将河中的小洲与堤岸连接起来。干水渠与长鸣河的功能相当于生态湿地，可以净化雨水，保证河流生态系统的正常运行。护城河的南岸建有河畔公园，北岸则建有步行街与建筑群。该项目的东部和北部有两条较浅的溪流，潺潺的溪水延续着"护城河"这一视觉主题，同时也起到排水的作用。

2

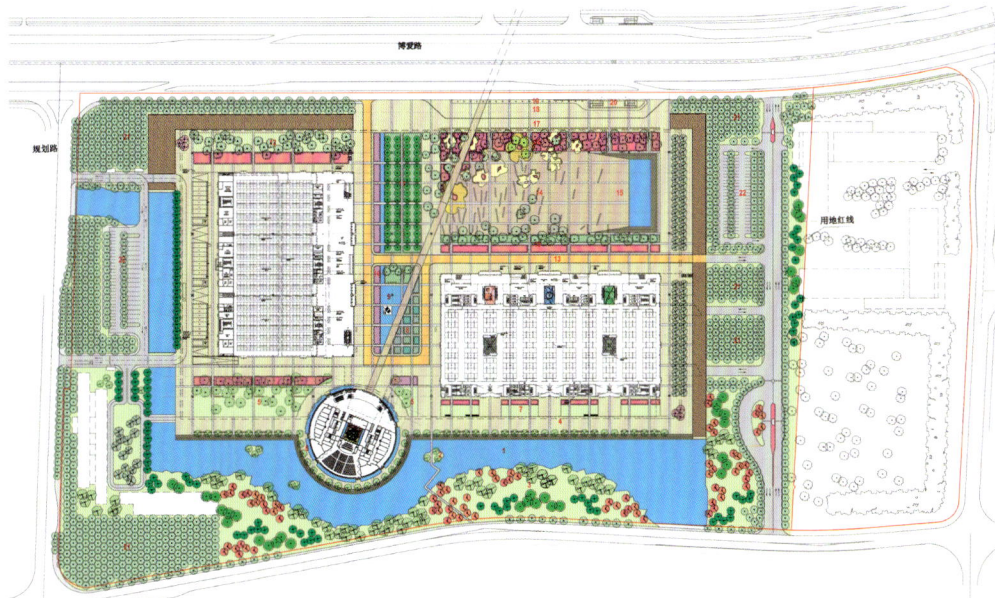

1 ECOLOGY CREEK WITH BIO-SWALE	生态溪涧	9 CENTRAL WATER FEATURE	中央渠带	17 ENTRY PLAZA	入口广场
2 LANDSCAPE ISLAND	景观小岛	10 SEASONAL COLOR	四季草花	18 DROP OFF	接驳区
3 WATER FRONT PARK	水岸公园	11 ENTRY LINEAR PARK WITH TREE BOSQUE	入口树阵公园绿带	19 FLAG POLES	项目旗帜
4 RIVER FRONT PROMENADE	河滨步道	12 NORTHSIDE GARDEN	北侧花园	20 TUNNEL EXIT	人行隧道出入口
5 SOUTHSIDE GARDENS	南侧花园	13 BOLLARD AND DRIVEWAY	车阻与车道	21 PERIMETER FOREST	周边树林
6 REFLECTION POOL	建筑映射池	14 MULTI-USE CENTRAL PLAZA	多用途中央广场	22 PARKING	停车场
7 YARD GARDENS	展厅花园	15 AMPHITHEATER / FOUNTAIN	户外水景圆场		
8 SEASONAL COLOR GARDENS	季节性彩缤花园	16 SIDE GARDENS	广场侧边花园		

总平面图

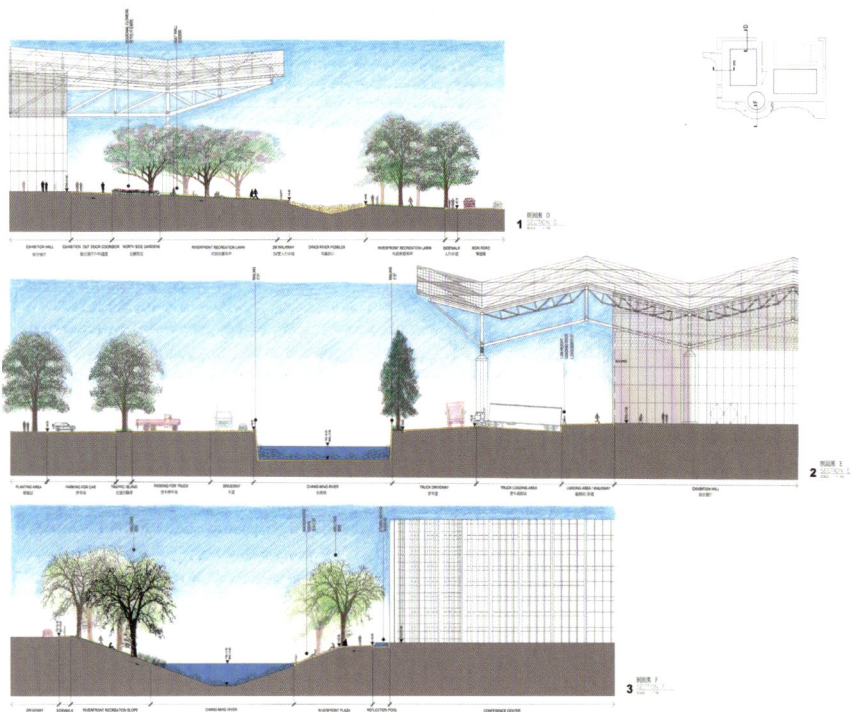

植物景观剖面图 1

1　林阴路

2、3　入口处

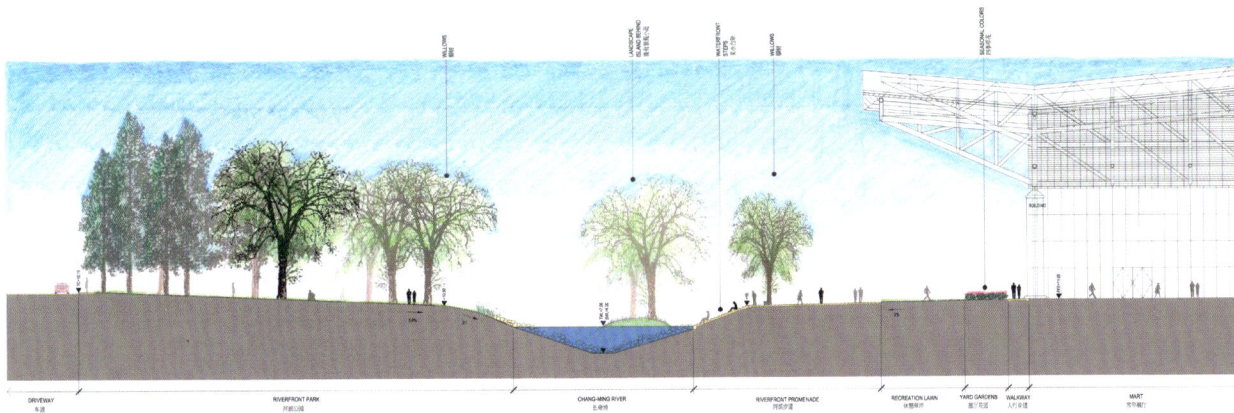

植物景观剖面图 2

The landscape design for the Zhongshan International Exhibition Center frames the Trade Mart, Exhibition Hall, and Conference Center buildings in a coherent, park-like setting that celebrate arrivals, provides a variety of gathering spaces, and complements the functions and design of the buildings.

The overall complex sits on a low podium, with east and west edges defined by a deep, layered "forest band" of alternating rows of trees. Two major landscape elements—an Entry Exhibition Plaza serving VIP arrivals and a major Public Plaza along Boai Road—create distinct open space zones that connect and provide focus for the building complex, which is in turn surrounded by a "moat" or water system consisting of Chang-Ming River and dry channels.

In the northeast sector is the Public Plaza, defined by the roof structure and providing a large, flexible venue for performances and outdoor exhibitions and shows. The eastern portion of the Public Plaza slopes gently to create

an amphitheater bounded at the east edge by a rectilinear water element.

Encircling this building complex is the shallow "moat" or river/bioswale system that provides an aesthetic feature while also collecting runoff from the building roofs and hardscape. Along the south and west edge, the Chang-Ming River will be a dry creek lined with rock mulch in the dry season and will accommodate storm water during the rainy months, with a small island connected by pedestrian walks. All the storm water runoff will be collected and drained to the dry channels and dry creek which function as a bio-swale and allow cleaner water returning to the river system. The south edge of the river is informally landscaped to create a riverfront park, while the north edge against the buildings serves as a more formal promenade. Two shallow dry creeks continue the visual motif of the "moat" to the east and north, also providing for drainage as needed.

1 中心水景
2 小溪
3 护城河
4 入口处
5 花园

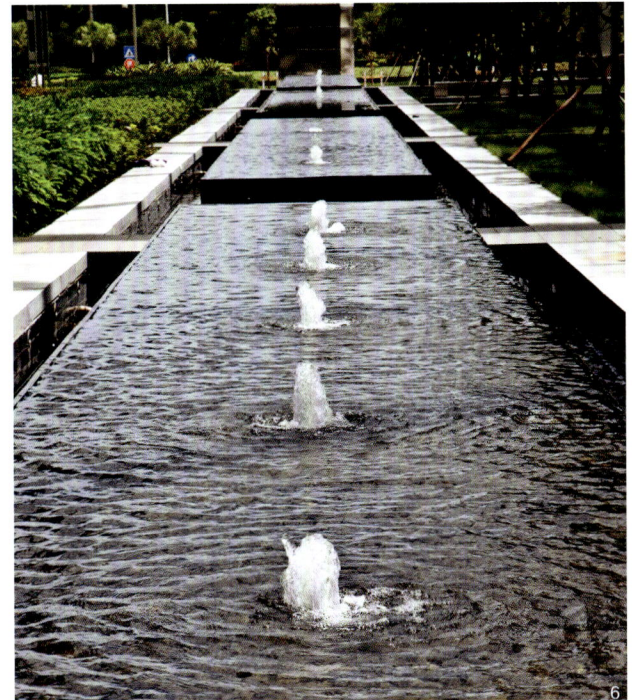

静穆的空间 —— 兰格蒂科墓园

Solemn Space — Cemetery at Langedijk

撰文 / 图片提供：Karres en Brands 景观设计事务所　　翻译：刘宏阳

由于兰格蒂科现有的墓园不具备扩展的条件，因此政府委托景观设计师设计了一处占地 60 000m² 的新墓园。该墓园选址于荷兰的 Zuid Scharwoude，毗邻一处包含公园、游乐场、小型公园和其他设施的休闲区域。水元素是该场地的一大特色，场地内有一座大型池塘和一条长长的线性水道（其地表水位高低不等），连同围海造田的排水管道和周围的运河，构成了该项目的结构框架。

该项目旨在设计一座包括 2500 座成人墓穴和 180 座儿童墓穴的公共墓园。一期工程现已竣工，建成了 1200 座成人墓穴、90 座儿童墓穴、等候区、管理区、骨灰盒纪念柱以及撒放骨灰的纪念花园。

荷兰的北部属于多风性气候，而私密性和庇护性在该地区的墓园设计中是十分重要的。因此，设计师决定将整座墓园分成一片片墓区，各墓区被高高的树篱环抱。墓园内的公共用地网络始于墓园的入口处，继而贯通整座墓园，其不仅连接着一片片墓区，还连通着墓园与毗邻的休闲区域。从视觉角度看，公共用地网络在公共用地与宽广的周边环境之间形成了视觉的连贯性。

该墓园在现有的殡仪馆附近兴建了一座大型停车场，周围环绕着种满绿草和树木的公共绿地，而且停车场与墓园之间设有直接通道。景观设计师与建筑设计师通力合作，将墓园现行的入口与设有屋顶的等候区域融为一体，这也是大坝工程中的一部分。另外，设计师还设计了多种多样的小型等候区域和坐椅区域。

由于地下水位较高，而墓穴需要埋在比地下水位深三倍的位置，因此墓园需高于周边的其他元素。其高差则通过延展的斜坡进行调整，这座斜坡贯穿整座墓园，并且毗邻线性水道。维护墓园的建筑则建在了地平面高度，而非建在墓园高起的位置，这样只有一

① 太平间　　　　　⑨ 钟楼
② 停车场　　　　　⑩ 骨灰纪念柱
③ 入口建筑　　　　⑪ 植被
④ 儿童墓地　　　　⑫ 树篱
⑤ 成人墓地　　　　⑬ 待建墓园
⑥ 维护墓园的建筑　⑭ 露台
⑦ 道路系统　　　　⑮ 桥梁
⑧ 火山灰原

小部分建筑能够进入人们的视野。

墓园内的墓穴被分布在 8 个不同的墓区，由高高的树篱环抱，以此保证其私密性。每片墓区的布局和环境各不相同，墓区内的树木和植被也各具特色，其中墓碑的安放与设置均因地制宜，与周边环境相协调。

设计师在儿童墓穴的设计上投入了更多的心血，其布局也十分独特。儿童墓穴被分布在两个不同的墓区，但其安葬的位置却位于同一地下深度。

青翠的草坪与稀疏种植的树木形成的公共用地网络在整体上略低于墓地，连接着墓园与其周边的环境，同时也连接着各处的私人墓区。公共用地网络与更广阔的道路系统之间的连接清晰可辨，因此墓园被牢固的镶嵌在其周边的景观之中，并且与休闲区域完美地融为一体。设计师还在墓园中增设了一些特别的设施，如通往水边的台阶和作为观景点的平台，从这些地方眺望，便能观赏到如茵的草坪和葱郁的树木。沿着公共用地网络分布着等候区、骨灰盒纪念柱和纪念花园。

该墓园的道路系统内生长着很多桦树，这些树木将整座墓园有机地统一起来，强化了公共用地空间的氛围。

As there were few options within Langedijk for extending the existing cemetery, Karres en Brands landscape designers were asked to design a single municipal cemetery covering an area of roughly six hectares. A suitable site was found in Zuid Scharwoude, the Netherlands. The new cemetery adjoins a recreation area with a park, playing fields, allotments and other facilities. Water is a particular feature of the area, a large pond and long, linear watercourses at various surface levels provide structure for the surroundings. Like the polder drainage ditch and neighbouring canal, these anchor the site.

The brief was for the design of a public cemetery for 2500 graves and 180 child graves, with the completion of 1200 graves and 90 child graves, a waiting area, management area, urn columns, and a garden of remembrance where ashes can be scattered in the first phase.

When designing a cemetery in the windswept province of North-Holland, it is very important to provide a sense of intimacy and shelter. We therefore decided to divide the cemetery into burial fields ringed with high hedges. These are connected by means of a parkland network that lead through the cemetery from the entrance, simultaneously providing connections to the neighbouring recreation area. Sightlines through the network link the parkland environment with its wider surroundings.

A large car park was made next to the existing mortuary, which is also surrounded by parkland with grass and trees. There is a direct link between car park and cemetery. In collaboration with Onix, the design for the actual entrance to the cemetery was combined with the design for the roofed waiting area, which is of one continuous piece with the dam. The design also comprises a variety of smaller waiting and sitting areas.

Because of the high water table, and because graves are layered three deep, the cemetery is now higher than its surroundings. The height difference is spanned by a continuous slope that runs round the entire cemetery and abuts the watercourses. The maintenance building is not sited in the raised part of the cemetery, but at the current surface level, where it is largely out of sight.

The graves are distributed over eight burial fields, and are given intimacy by the high hedges that enclose them. Each burial field has its own layout and atmosphere, with its own vegetation and trees, and with the gravestones being placed in a way particular to the area in question.

Particular attention has been given to the child graves, which have their own layout. They have been distributed over two areas and are dug to a single depth.

The parkland network connects the cemetery with its surroundings; it also connects the individual burial fields. Thanks to its physical connections with the wider system of paths, the cemetery is anchored within the surrounding landscape, and also dovetails well with the recreation area. At all points, the network created by the lawns and widely spaced trees lies slightly below the burial fields, it reveals itself to its surroundings, when special features are added, such as steps down to the water, or a terrace that serves as a viewpoint. Lying along this network are various points, such as a waiting area, a urn columns and a garden of remembrance.

The cemetery is unified by birch trees, standing largely in the system of paths, these strengthen the parkland atmosphere.

终年不败的绿叶 —— 昆西庭院

Year-round Leaves — Quincy Court

撰文：Julie D. Taylor　　图片提供：Scott Shigley　　翻译：刘丹春

芝加哥南州大街旁边的一条小街如今正经历着美丽的蜕变，这要归功于 Rios Clementi Hale 工作室设计的有着春天气息的雕塑群。这个采用多学科设计方法的设计公司将昆西庭院从一个旧市区街道的残余部分转变为被芝加哥市民和游客津津乐道的迷人场所。

洛杉矶的设计师们通过大胆的图样设计为这半个街区的空间设计了遮阳结构、多样的坐椅布局，并对硬质景观进行了改善。昆西庭院以及其毗邻地产由美国总务管理局（GSA）购得，以增强安全性和促进联邦政府规划的芝加哥商业区未来的发展，其更直接的目标则是提升此处的市容市貌，使其对公众更具吸引力。

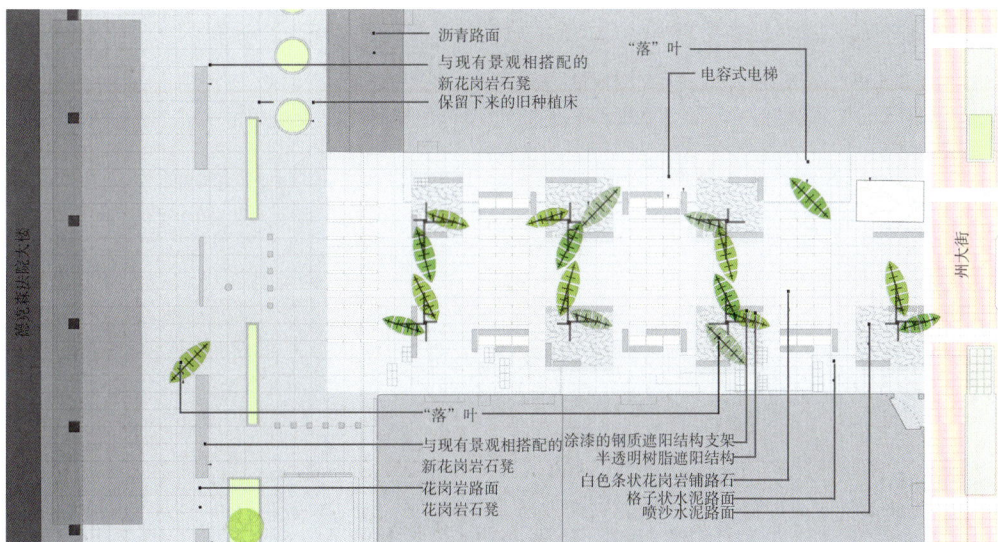

沥青路面
与现有景观相搭配的新花岗岩石凳
保留下来的旧种植床
"落"叶
电容式电梯

德克森法院大楼

州大街

"落"叶
与现有景观相搭配的新花岗岩石凳
花岗岩路面
花岗岩石凳
涂漆的钢质遮阳结构支架
半透明树脂遮阳结构
白色条状花岗岩铺路石
格子状水泥路面
喷沙水泥路面

昆西庭院东临州大街，西临德克森联邦办公大楼（密斯·凡德罗设计，1964年）、北临南州大街220号消费者大厦（Mundie & Jensen设计，1913年）、南临南州大街230号（原址为班森－里克森百货商店，阿尔弗雷德·S.·阿舒勒设计，1937年）、西杰克逊大道110号跨联盟大厦后部（A. Epstein & Sons国际设计，1961年）。设计元素——抽象的树形、半透明的桌子与组合照明、白色花岗岩铺路石，为联邦广场上巨大的现代建筑和历史悠久的州大街的步行区之间提供过渡。

"遮阳结构以及硬质景观的细节作为昆西庭院特色的构成元素展示着这个地方的设计，"Rios Clementi Hale工作室的项目设计师Jennifer Cosgrove（美国建筑师协会会员）指出："设计灵感来自于在城市中以及联邦大学到处可见的皂荚树、芝加哥传统建筑所用的白色陶土、现代广场的密斯式栅栏和德克森联邦办公大楼的外立面。"

Rios Clementi Hale工作室——美国总务管理局（GSA）为全国"第一印象项目"挑选的两个景观设计公司之一——因将未充分利用区域改造为美丽且游人众多的公众空间而闻名。这家公司成功地将加利福尼亚州格伦代尔的一条小巷转变为一个十分艺术化的象棋公园，赋予这条街以生命，创造了一个安全的公众场所，并且改善了城市结构。"关于成功的公共空间设计，我们的目标是提供社区集会、公众交谈以及娱乐的空间。"Jennifer Cosgrove如此说道。

1　新长凳为行人提供了更多的休息空间
2　水泥长凳和闪光树脂桌为人们提供了舒适的午餐地点
3　融入到硬质景观中的坐椅鼓励了休闲、公共聚会

2

新广场上有七个钢质的树形遮阳结构和三个入夜后在广场上空点亮的半透明亚克力板。"树"的根部固定于树叶形的喷砂混凝土中。新的花岗岩长凳和铺路石与现有坐椅和硬质景观并存，而这些新建的公共设施的设计语言通过混凝土长凳和内部发光的半透明树脂桌展现出来。地面镶有四只巨形树叶，看上去像被"风城"（芝加哥别称"风城"）的强风吹落后散落在地上一般。

Rios Clementi Hale 工作室成立于 1985 年，人才济济。此次非凡的实践为他们带来了具有协作精神和领域多元化的国际声誉，并通过史无前例的多学科设计而屡获殊荣。为表彰其各方面的工作，美国建筑师协会加州理事会将 2007 年公司奖颁发给该工作室，这个奖是给予一个团队的最高荣誉。

3

A former side street just off South State Street in Chicago is getting a beautiful makeover, thanks to a spring-like sculptural grove designed by Rios Clementi Hale Studios. The multi-disciplinary design firm transformed Quincy Court, a remnant of an old downtown street, into an engaging gathering place for Chicagoans and visitors.

Using bold graphic forms, the Los Angeles-based designers provide canopy elements, a variety of seating configurations, and hardscape improvements to the half-block space. Quincy Court and adjoining properties were acquired by the U.S. General Services Administration (GSA) to provide added security and future expansion for the federal government's downtown Chicago campus. The more immediate goal was to upgrade the look and feeling of the space to make it more appealing for public use.

Quincy Court is bordered by State Street to the east, Dirksen Federal Building (Mies van der Rohe, 1964) to the west, 220 South State Street/Consumers Building (Mundie & Jensen, 1913) to the north, and 230 South State Street (originally Benson-Rixon Department Store, Alfred S. Alschuler, 1937) and the back side of 110 West Jackson Boulevard/Transunion Building (A. Epstein & Sons International, 1961) to the south. The design elements—abstracted tree forms, translucent tables with integrated lighting, white granite accent pavers—provide transitional scale between the monumental modern architecture of Federal Plaza and the pedestrian scale of historic State Street.

"The canopy elements and hardscape details tell the story of the site by alluding to the elements that form Quincy Court's character," notes Rios Clementi Hale Studios project architect Jennifer Cosgrove, AIA. "The design is inspired by the honey locust trees used throughout the federal campus and prevalent in the City, the white terra-cotta detailing of historic Chicago buildings, the Miesian grid of the modernist plaza, and the reflected light patterns of the Dirksen Federal Building façade."

Rios Clementi Hale Studios—one of two landscape architecture firms selected by GSA for the nationwide First Impressions Program—is well-known for turning underused areas into beautiful and well-traveled public spaces. The firm's success at transforming an alleyway in Glendale, CA, into an artful Chess Park has enlivened the street, created safe public gathering space, and improved the city's civic fabric. "Our common goal for the design of successful public space is to accommodate community gathering, public discourse, and enjoyment for people," says Cosgrove.

The new plaza features a series of seven tree-like canopy elements made of steel and three tones of translucent acrylic panels that are lit from above after dark. The "trees" are rooted by sandblasted concrete in an abstracted leaf pattern. New granite benches and pavers join existing seating and hardscape materials, while a new site furniture language is introduced using concrete benches and translucent resin tables glowing with inner lighting. Four large leaves are situated on the ground, seemingly scattered on the pavement, the "result" of a strong gust well known to the Windy City.

Rios Clementi Hale Studios encompasses myriad talents in one firm. Established in 1985, this extraordinary practice has developed an international reputation for its collaborative and multi-disciplinary approach, establishing an award-winning tradition across an unprecedented range of design disciplines. Acknowledging the firm's varied body of work, the American Institute of Architects California Council gave Rios Clementi Hale Studios its 2007 Firm Award, the organization's highest honor. The architecture, landscape architecture, planning, urban, interior, exhibit, graphic, and product designers at Rios Clementi Hale Studios create buildings, places, and products that are thoughtful, effective, and beautiful.

城中 "起居室"

Urban "Sitting Room"

撰文：Diamond Architects　　图片提供：Step Haiselden　　翻译：王玲

1　休闲坐椅 1
2　人们在此休息闲谈
3　"起居室" 周围的植物

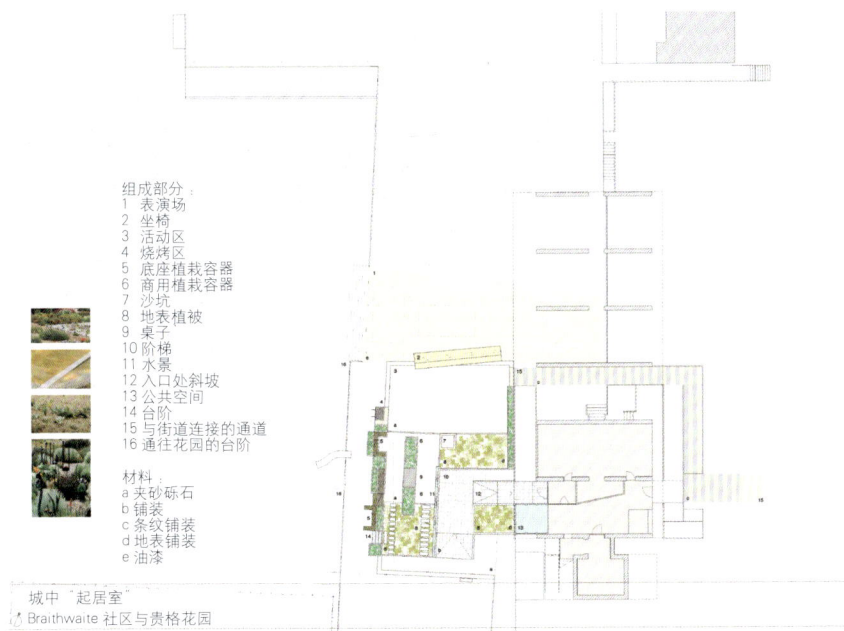

组成部分：
1 表演场
2 坐椅
3 活动区
4 烧烤区
5 底座植栽容器
6 商用植栽容器
7 沙坑
8 地表植被
9 桌子
10 阶梯
11 水景
12 入口处斜坡
13 公共空间
14 台阶
15 与街道连接的通道
16 通往花园的台阶

材料：
a 夹砂砾石
b 铺装
c 条纹铺装
d 地表铺装
e 油漆

城中"起居室"
Braithwaite 社区与贵格花园

总平面图

这是一个小型的设计项目，它使伦敦一处20世纪60年代的柏油屋顶摇身一变成为惠及当地居民的城市露天"起居室"。

Diamond Architects建筑设计事务所、Public Works艺术与建筑设计集团与当地居民共同商讨关于这块位于Banner Estate的户外公共空间的设计方案。项目原址是该住宅区停车场上的一个开阔的架高露台，后面是一座19层的住宅楼，周边则是低层住宅。

设计师充分考虑到当地居民的想法和建议，并让他们积极深入地参与到项目的设计过程中。设计师力图打造一个能够吸引公众、提升建筑公共价值的可持续性设计，因此露台秉承其原始用途，并有效地促进了当地社区各项活动的开展。

该项目的预算有限，而且露台混凝土板有限的承载力也是设计需要克服的一大难点。设计包括一系列遵循当地居民意愿及公共空间设计理念的措施和铺装，使社区和公共空间有效地结合在

一起——通过景观小品、游乐设施、植被以及铺装为当地居民营造出活动、娱乐、交际和沉思的空间。

"起居室"的混凝土扶手椅和带图案的"地毯"吸引着人们在此流连忘返——人们或三五成群地到这里野餐，有时上班族也会在这里进行露天午餐。休闲坐椅不仅为人们提供了观景的舒适场所，而且当孩子们在球场或下面的游乐区嬉戏玩耍时，家长们还可以坐在这里悠闲地聊天。

地面上彩绘的图案不仅与附近建筑物立面上的窗户相得益彰，同时也充当了如"跳格子"等一些经典铺装游戏的功能。Public Works艺术与建筑设计集团在地面图案的施工期间，现场添加

了一些为当地儿童设计的元素——富有趣味性的"自然本色"图案彰显出孩子们心目中对花园的想法；户外的乒乓球台也为露台平添了些许休闲娱乐的味道。

架高露台上的景观设计与后面街道上的开放空间巧妙地融合在一起，如球场、儿童游乐场以及在成熟法国梧桐树掩映下颇具野趣的贵格花园。成排的植被为曾被忽视的大型种植槽增添了无限生机，与周围的绿色空间浑然一体。当地多风的环境要求植被具有较强的生命力，因此设计师选择适宜在海边和沼地生长的典型的本土耐旱植被，如草地、海冬青、一些野生植物和草本植物等，使社区居民沉浸在大自然的怀抱中，尽

情地感受田园般的生活气息。

新型照明系统也被用于住宅楼的设计中，它们照亮住宅楼下面的开放空间，增加了空间的有效利用性；低层的城市照明系统增强了露台夜间的使用功能。

1、2　混凝土扶手椅和带图案的"地毯"

3、4　地面铺装1

5　室外乒乓球台

The Podium Project is a small-scale design intervention that has transformed a bare expanse of asphalt on a 1960s London housing estate into an open-air urban "sitting room" that benefits local residents and office workers.

Diamond Architects in collaboration with Public Works, an art/architecture collective, were commissioned to undertake a series of workshops with local residents in order to develop the brief and design for this piece of shared public outdoor space within the Banner Estate—a large raised open podium over a residents' car park, in front of a 19-storey residential tower block and adjacent to a low-rise housing development.

The process of developing the scheme was the result of intensive engagement with the local residents to develop a brief from their desires. The design principal was to develop a sustainable scheme to enhance the building's public realm that would engage the residents, so that the podium maintains itself through use and act as a catalyst for local community activities.

The scheme had a very modest budget and had to be developed in the context of serious loading issues which were found to exist with the podium's concrete deck. The design consists of a series of interventions and surface treatments, relating to resident desires and the idea of a public space, which binds the community and links to the public uses around it. By introducing pieces of furniture, games, planting and surface treatments, the project has created areas for activity, play, interaction as well as contemplation, for residents and the wider community.

A "sitting room" of concrete armchairs and ground patterned "rugs" attracts residents to linger and picnic, and is also used by local workers for al fresco lunching. Informal bench seating provide places from which to view the estate and for parents to chat while their children play on the ball court and play area below.

The painted ground patterns reflect the repetitive fenestration on the tower façade and also act as guides for classic pavement games. The painting scheme includes design elements that were developed by resident children at on-site projects faciliated by Public Works during the construction. The playful interference of "natural" ground patterns relates to their ideas of gardens, culminating in the

door to an imagined secret garden. An outdoor table tennis table adds to the Podium's play potential.

The landscape scheme on the raised podium binds together the adjacent open spaces at street level behind—the ball court, childrens play area, and the peaceful, slightly wild nature of the Quaker Gardens with its existing mature plane trees. The planting brings to life previously neglected large existing planters with a line of trees to link the surrounding green spaces. The planting in this windswept environment is robust and wild, using plants typical of warm dry marginal native landscapes—sea edge and moor. These include grasses, sea hollies, some wild planting, and herbs, which will be used to introduce the community to the sensory nature of plants and food growing.

New lighting has been introduced around the tower; lighting the open spaces under the tower helps define the space and low level urban lighting encourages evening use of the podium.

1　休闲坐椅 2
2、5　地面铺装 2
3　建在停车场屋顶上的架高露台
4　耐干旱的本地植物

滨水空间

Waterfr

nnt Space

德国布莱梅海港重建后的公共开放空间

Public Open Space in a Re-animated Harbour Site, Bremerhaven, Germany

撰文 / 图片提供：Latz+Partners Landschaftsarchitekten　Emporis 网络编辑 Pedro F Marcelino　翻译：刘建明

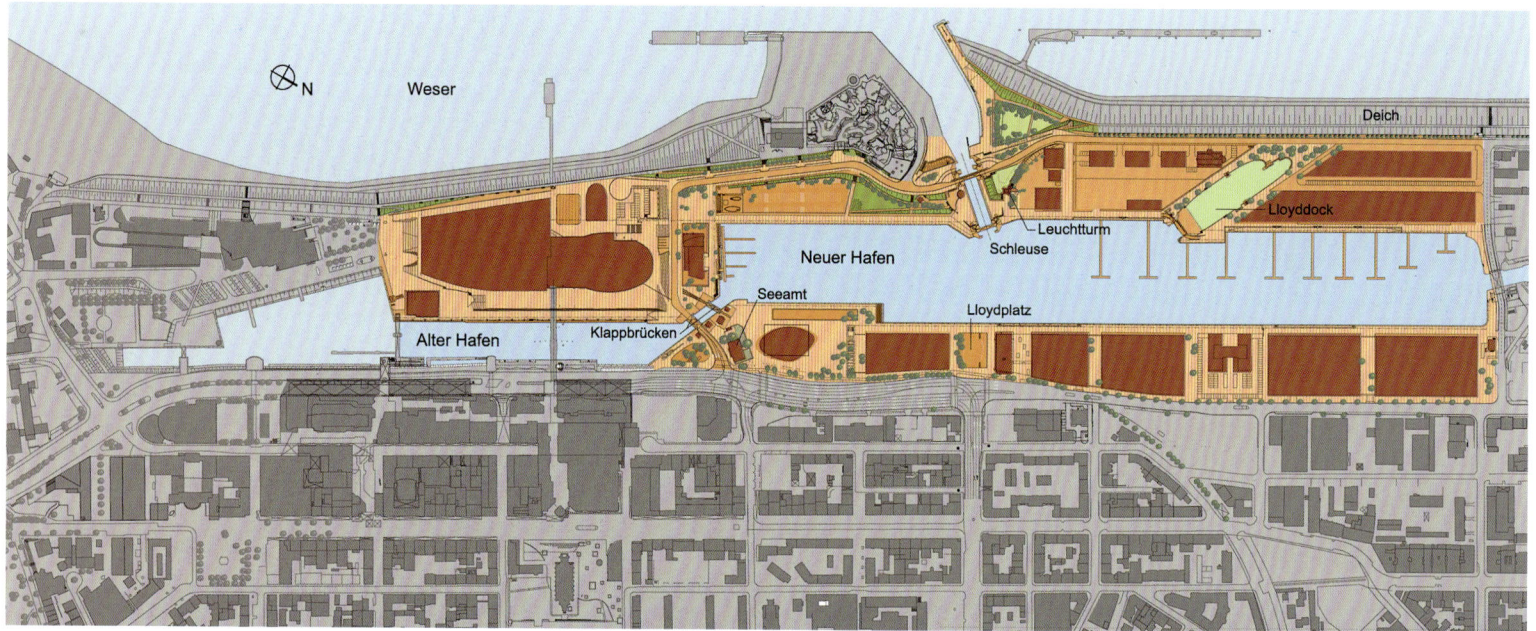

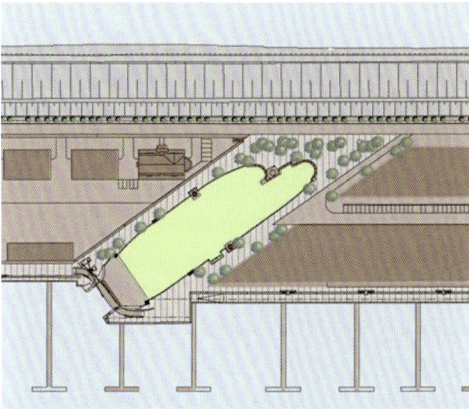

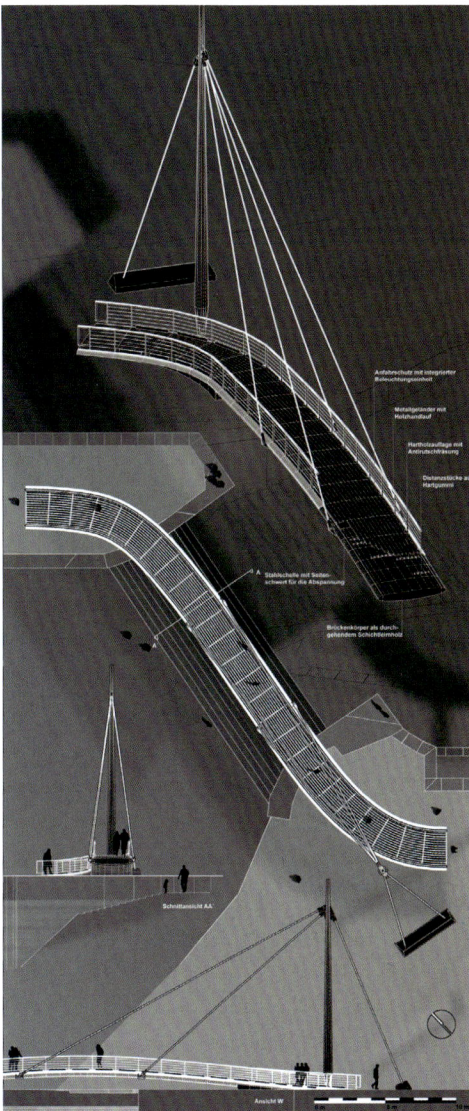

1999 年，市政当局决定重建日益衰败的港口，在城市中心建立一个包含公共开放空间、步行道、文化景观在内的空间，打造一个集居住、商业和娱乐等功能为一体的全新场所，改造现有的城市结构类型。

2001 年，Latz + Partner 和 Latz·Riehl·Partner 公司被委以重任，并承担重建城市公共开放空间的设计任务，负责各个功能区域的统一规划，希望通过设计能为港口居民重新找回滨水生活的感觉，再次证明布莱梅市独特的城市定位，使其城市公共开放空间的定位与历史紧密相连。

当"2005 年布莱梅国际帆船展"举办时，全球闻名的帆船纷纷云集于"古老的新港"，此时的改造工程已完成了三分之二，布莱梅市民可以骄傲地将崭新的市容展现给来访的 170 万名游客。

规划与实现

该项目的规划在遵循传统设计元素的基础上稍加改动，沿用现有的城市布局以保证长远的使用和发展，以城市的气候条件作为对城市定位的积极因素，旨在创建一种能够满足各种需要的城市布局形式。

　　公共设施是这个项目中不可或缺的一部分，很多公共设施都是为该项目特制的，如排水设施、高压输电线以及与电话配套的服务设备。垃圾桶、形态各异的长椅和路灯使城市街道焕然一新，同时也突出了码头的特色。

　　地面的铺装材料是当地最常见的，也是最能代表码头、步行道和广场特色的铺装材料——地毯式花岗岩。由于这里相继出现的建筑物风格迥异，设计采用石灰石和沙岩营造出一个静谧而宽广的空间。而只有在个别场所，地面结构才会有所不同，例如设计师采用阶梯式的镶嵌来突出一些独特的场所。海港的入口正是采用了这种阶梯式的镶嵌方法，突出了其从城市、河流到古老灯塔这一沿途风景线中的核心位置。

　　不同的灯光设计增强了空间感和视觉感。多功能柱子顶部的蓝色灯光不仅可以指引方向，更清晰地勾勒出了建筑群的轮廓。

　　稀疏栽种的树木减弱了风势，并与各种公共设施共同展现出由一个纯粹的港口城市向充满活力的城市的转变。由于保留了造船和航海等历史产业，陈旧的工业场所因其开放性和功能性而散发出全新的魅力，新港口移除了曾经不具有任何象征性的一切设施。"古老的新港"逐渐发展成为码头和城市中心之间的核心区域，布莱梅市民和无数的游客也纷至沓来。

In 1999 the town established the BEAN - Development Company Alter · Neuer Hafen GmbH. The task was to re-animate the devastated site and to create a new quarter near the city centre with public places and promenades, with residential, commercial and recreation areas, with tourist and cultural attractions—thus initiating a structural change of the whole town.

In 2001 Latz + Partner and Latz · Riehl · Partner were commissioned to integrate the different projects envisioned in a master plan and to develop the public open spaces for the new town quarter. The aim was to find again the town's identity by re-conquering the waterfront for its inhabitants, the new uses tightly connected with the experience of its history. The realization of the project was and is only possible by tight co-operation of the different disciplines involved.

At the "Sail Bremerhaven 2005" the world's most famous sailing ships gathered in the Old · New Harbour, two thirds of its new face already realized and proudly presented by the town to 1.7 millions of visitors.

Planning and Realization

The planning follows the strategy of a metamorphosis, which is based on traditional elements. It refers to the existing urban pattern to guarantee a long-term use and development. It works with existing climatic conditions as a positive identity. It aims at establishing a strong language of forms for manifold demands.

It works with equipment, developed especially for this site: Drainage elements, service units for high voltage power lines and telephone, litter bins and several types of benches and lamps form a program for street furnishing and emphasize the specific character of the quays.

The planning works with surface materials common in the place. Thus one material characterizes the quays, the promenades and squares: continuous carpets of granite sets, limestone and sandstone create quiet and generosity in view of the future heterogeneous built up areas. The structure of the surfaces is changing only at special places. Like inlays or "treads" particular materials are marking exceptional spots. The wooden deck of the Lloyd Place is one of these inlays representing the entrance to the harbour and forming the centre of an important sightline from the city to the river and the historic lighthouse.

Spatial impression and visual connections are supported by a differentiated light concept. The blue lights on top of the new multifunctional poles give orientation from far away and mark already now the silhouette of future buildings.

Sparsely inserted trees are breaking the wind. Together with the equipment they illustrate the change from a pure harbour to a vivid part of the town. The new image of the former industrial site, its publicity and usability by keeping at the same time the shipbuilding and seafaring heritage, has removed symbolic and physical barriers. Having become the core of the new town quarter between the dikes and the city centre, the Old · New Harbour is visited and used with pleasure by the citizens of Bremerhaven and numerous tourists throughout the year.

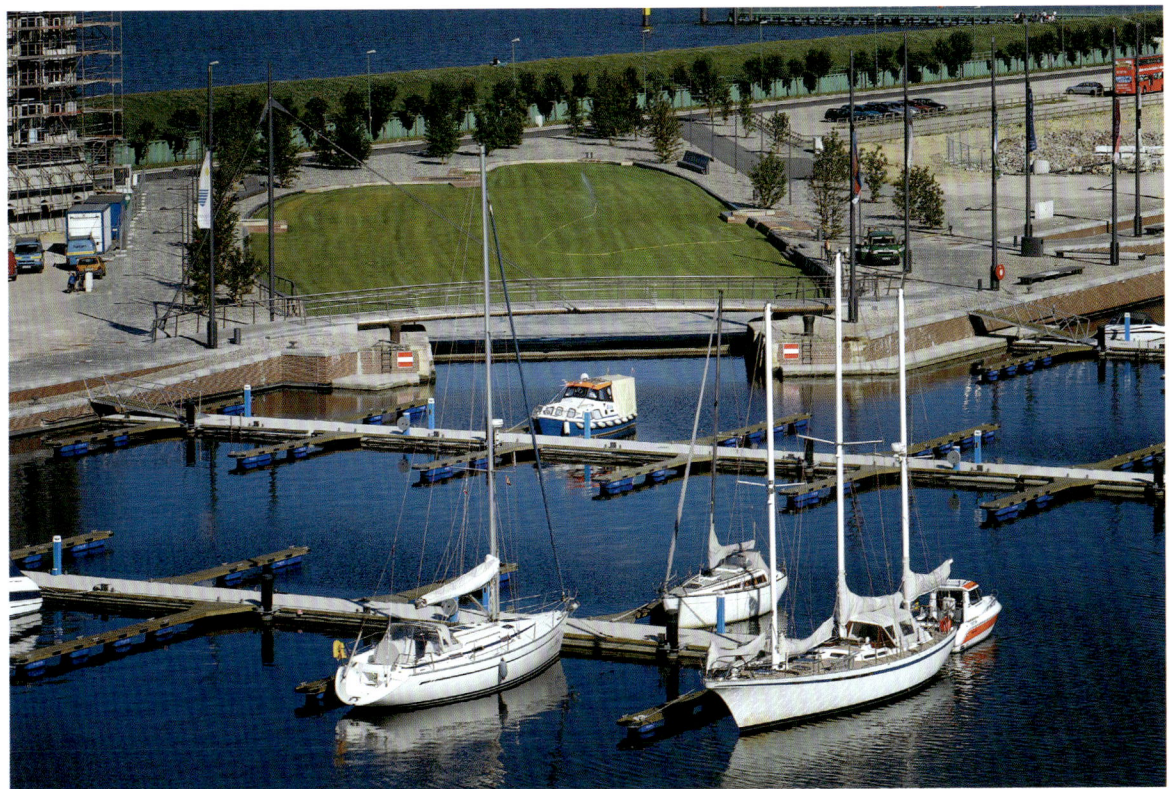

再现安大略湖的湖畔魅力 —— 多伦多H$_T$O 滨水公园

Rediscover the Waterfront — Toronto H$_T$O Park

撰文：Emporis 网络编辑 Pedro F Marcelino

图片提供：JRALA Claude Cormier Architectes Paysagistes Neil Fox Pedro F Marcelino　　翻译：刘建明

1　俯瞰 H$_T$O 公园
2　选址规划示意图
3、5　多伦多市区的卫星航拍图
4　枫叶码头东侧 3D 图（竞标阶段的）

1

WATER MARKS

±1.5
±2.52
±10mm EXTENSION
TOPOGRAPHY

8%
±1.8
±2.5%

DUNES

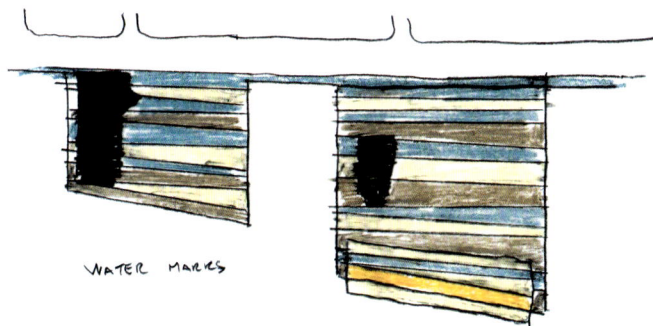

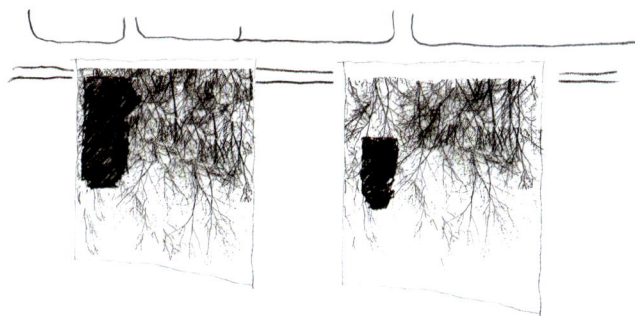

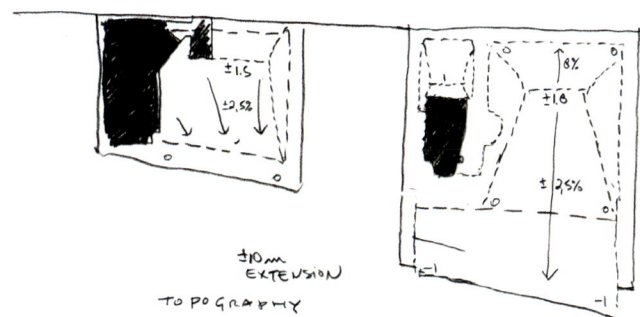

地形、植被、水域标志和绿岛手绘草图

Bremner Boulevard
Mariners Tier
Lake Shore Boulevard West
Rees Street
Gardiner Expressway
Queen Quay West

众多国际性城市的评定标准都将多伦多市列为全球最适宜居住的城市之一，而多伦多的城市形象也完全符合这一殊荣。作为带动加拿大经济发展的领军城市，有260万常住人口的多伦多市被公认为北美地区绿化范围最广、最具经济活力和最具创新意义的城市。

城市海滩已成为大多数西部海岸城市甚至内陆城市的公共空间，H_2O公园证实了即便是面积最为狭小普通的项目中亦有可创新的空间这一道理。更值得一提的是最近几十年来多伦多市民几乎忽略了安大略湖的存在，但是H_2O公园项目的巨大成功似乎又再一次唤起了人们对湖畔的美好回忆，并再次引起了市民对安大略湖的热切关注。多伦多城区的主干道Gardiner高速公路南侧一带出现了许多新的公共区域，这些公众空间迅速发展成为社区的主要休闲之所。近些年的极端气候条件改变了滨水地带的居民的消暑方式，这就更加突出了市政当局逐步探求改善滨水景观环境的迫切需要。

该项目选址于Yonge和Bathurst大道之间的核心

腹地，项目落成后很快就成为全城 1470 处景观设施中最具标志性的新型绿色景观项目。H₂O 公园不仅是一个备受市民喜爱的景观项目，还引起了国际上对它的广泛关注，迅速成为当地最受欢迎的场所。凭借备受瞩目的景观和环境特色，H₂O 公园给游客带来了独特的视觉感受。

设计团队早在设计之初就预料到了会产生这样的效应，他们从 Georges Seurat 的代表作《大碗岛的星期日下午》(1884 年～1886 年) 中获得了设计灵感，用一种融合了色彩、光影的现代设计手法重现了现实生活中一种平静而略显倦怠的冥想状态，随处可见的垂柳摇曳出万种风情。其独特的景观包括一条湖畔散步道、各种视听艺术装饰、太阳能遮阳伞和风力发电机、湖泊净化系统（据环境工程团队 Sustainable Edge 的解释，这是一种漂浮的生物修复岛屿，可以减轻因多雨天气带来的溢流问题，还可以带来视觉上的美感，并为野生动植物提供栖息地。）以及彼得街盆地的移植生物群落。

皇后码头西至安大略湖的剖面图

公园剖面图

1 茂盛的花园是滨水地区的点缀

2 灵感源泉——设计 Georges Seurat 的代表作《大碗岛的星期日下午》(1884～1886)

3 从枫叶码头东面的绿岛观看到的沙滩地带

4 H₂O 公园吸引了众多游客

项目基地曾饱受环境污染，因此，可持续体系成为该项目设计的首选方案。Garrison 河的河口湿地可以提供足够的水处理资源，采用 100% 可再生的能源，使公园的污水排放率几乎为零。受污染的土壤被圈起来，不用做垃圾掩埋，以避免扩大污染范围，如此圈护可以将公园的使用者和污染地有效地隔离。雨水经过渗透处理后被逐步流向公园四周，并可以在夏季利用湖水来灌溉，在公园的边缘地带设有专门的鱼类栖息地。

进入 H₂O 公园后，游客可以沿着绿色的小沙丘往山上走，沿途种有柳树和枫树；而下坡则通向一处沙滩，这里在夏季会设有黄色的太阳伞。码头的边缘有一条宽阔的散步道，微斜的坡面阻隔了噪音，公园内恬静的氛围使游客可以暂时远离城市中的喧嚣，而多伦多壮丽的城市轮廓线却如此之让人触手可及。整个项目的设计理念由地平面、绿色沙丘、精美的园艺、灯光、沙滩以及水平面六个层级分明、相互依托的元素构成，从而实现了由城市向水域的完美转换。

无论是步行、轮滑、骑自行车、泛舟、搭乘有轨电车还是自驾车，游客都可以方便自由地进出公园。地处市中心并靠近商业区的地理优势也促进了公园周边区域的发展，这样的理念正是源自生态学——不同的体系才能造就出多样性，正所谓娱乐与商业相辅相成，并进一步提升了市民的生活质量。

这个由 Rosenberg、Cormier 事务所设计的方案经常被引证为"多伦多市中心滨水地区发展的催化剂"的典型代表，H₂O 公园也成为首个湖滨地带的开发项目。在 H₂O 公园开放之日，当地的学生们将一个包含对多伦多滨水项目设想的"时空瓶"掩埋起来。这个"时空瓶"将在 2020 年的一个特别仪式上公布。或许，那时的多伦多已经成为与众不同的城市，正如 2007 年夏季多伦多最受欢迎的消遣去处 H₂O 一样，整个城市将布满一片阴凉。

Often considered by a variety of international urban indexes as one of world's best cities in which to reside, Toronto is finally claiming the visual responsibility attached to such label. The economic engine of Canada is home to a highly skilled population of 2.6 million and has increasingly been marketed as one of the greenest, most prosperous and innovative cities in North America.

While urban beaches are becoming common places in most coastal (and inland) western cities, H_2O Park proved that there is space for innovation in the smallest and most banal project. Perhaps more remarkably, after decades of little or no interest in Lake Ontario, Torontonians seem to finally have taken a fancy on their own lakeshore. That enthusiasm has resulted in the huge success of a magnificent design. New condominiums mushroom on the south side of the Gardiner Expressway, the city's main downtown highway, creating a new public an area that is quickly becoming a major recreation neighbourhood. Recent extreme weather patterns have also changed the way waterfront residents spend their summers, and this dictated the need for an enhanced waterfront that the city has been tackling step by step (albeit not always with the required energy).

3

4

1　公园周边环境鸟瞰图

2　H_2O 公园入口

3　无论是步行、骑自行车、驾驶汽车、乘坐有轨电车或是泛舟都可以随意游览 H_2O 公园

4　H_2O 公园成为欣赏安大略湖美景的最佳观景场所

In the heart of this intervention area, between Yonge and Bathurst Avenues, the H$_T$O project quickly became the newest flagship space of the City of Toronto's Parks, Forestry and Recreation Division, responsible for 1,470 other green areas. More than a pet-project, however, the park attracted international attention and quickly became a local favourite. With many noteworthy landscaping and environmental features, the park is also visually attractive and, yes, fundamentally... sexy.

The design team is certainly no stranger to this result.

Composed by Toronto's Janet Rosenberg from Janet Rosenberg + Associates in collaboration with Montreal-based powerhouse Claude Cormier, and Siamak Hariri from Hariri Pontarini Architects, the team drew some inspiration from Georges Seurat's famed painting a "A Sunday Afternoon on the Island of La Grande Jatte" (1884-1886). The sense of peace and languid contemplation was reproduced in real life with a modern twist on the colour, light and shadow games provided by scattered and somewhat romantic weeping willows.

Features include a lakeside promenade, various visual and audio art installations, photovoltaic parasols and urban wind turbine lampposts, "lake restorers" (explained by Sustainable Edge, the environmental engineering team, as "floating bioremediation islands to mitigate pollution from wet weather overflow while imparting aesthetic appeal and habitat for wildlife") and a naturalized biome in the Peter Street Basin.

The site's history is one of environmental damage. Sustainable systems were therefore a prerogative for the design. Wetlands at the Garrison Creek outfall provide adequate water treatment, and the park has a zero discharge rate, as well as a 100% renewable energy source. Contaminated soils were left undisturbed and capped, instead of being shipped to a landfill elsewhere, thus avoiding extending the damage. Capping allowed for distance between the park's users and this pollution. Rainfall is filtered through a set of pervious surfaces that infiltrate water and gradually disperse it after treatment. Irrigation in the summer uses exclusively lake water (as there often are irrigation bans in the city, due to high temperatures). Fish habitats were added on the edge of the park and in the slip made use of in-site recycled concrete.

As soon as visitors enter H_1O, they walk uphill through green dunes (or green islands re-enacting sand dunes) planted with willow trees and silver maples. The plane then slopes down to a sand strip with large, summery yellow parasols and a boardwalk at the very edge of the quay. The slight slope blocks noise off and the overall atmosphere of the site quickly induces the visitor to forget the city behind. Yet, Toronto's imposing skyline is literally meters away. The concept is composed of six superimposed and adjacent elements or layers (planes) that provide a transition from the city to the water: ground planes, green islands, expressive horticulture, lighting (by night), beach furniture and finally the water layer.

H_TO stands on two outer piers between Yonge and Bathurst Avenues, Maple Leaf Quay East and West, easily accessible by foot, rollerblades, bike, boat, tram or car, and providing easy accessibility for people with various physical impairments, all the way to the water's edge. Its city-centre location and closeness to commercial areas is prone to the development of an edge effect—the idea borrowed from ecology that the largest diversity happens where different systems meet: here, leisure will meet business, thus increasing quality of life.

This acclaimed Rosenberg/Cormier design has often been referred to as a "catalyst for Toronto's central waterfront development", being the first park in a series of projects that will green-up the lakefront. On opening day, a class of local school children buried a time capsule containing items and thoughts about the city's waterfront. This capsule will be dug up in a special ceremony in 2020. Perhaps by then Toronto will be a different city where a place in the shadow (or in the sun) is as much of a hot commodity as H_1O was during the summer of 2007—and others to come.

1   Gardiner 高速公路南侧的公共区域
2   在以城市轮廓线为背景的灯光设计和 CN 塔

马德罗港新建绿色地带与滨水地区复原

New Green Areas for Puerto Madero and Restoration of the Waterfront

撰文：Jimena Martignoni　　图片提供：Facundo de Zuviria　　翻译：刘建明

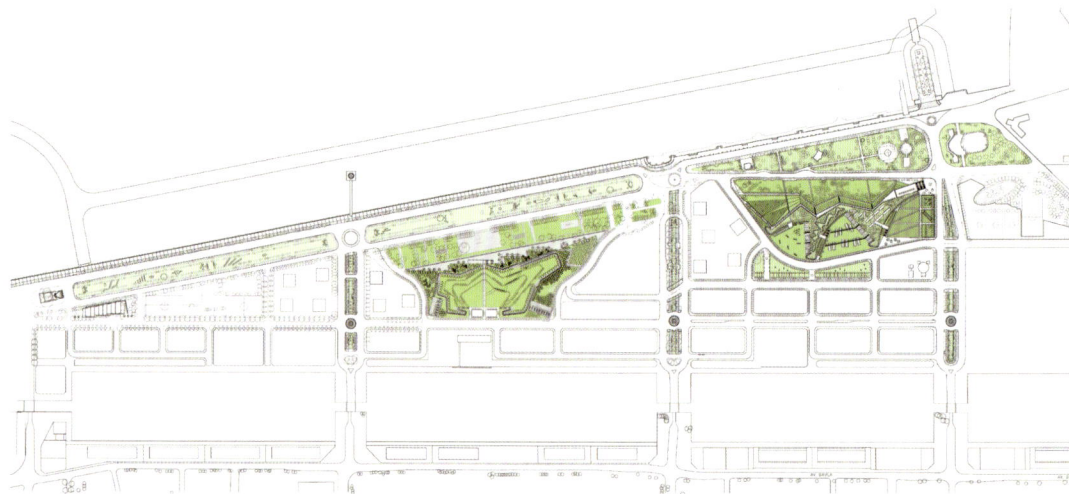

总平面图

1 Micaela Bastidas 公园的玫瑰花园
2 沿公园最高处而下的坡道，格宾外围植着地被植物和藤蔓

阿根廷首都布宜诺斯艾利斯是世界上最大的都会城市之一，由港口城市发展而来，逐渐沿河岸形成规模化的城市。

布宜诺斯艾利斯最初的港口于 1897 年竣工，名为马德罗港，25 年后新港的建设已势在必行，因为城市设施已经不能满足需要。马德罗港旧港口已被荒废，曾作为公共空间与海滨浴场的滨水地区在若干年后也将荒废。此后，有关复原重建旧港的提议从未间断过，但是一直没有得到采纳实施，直到 1989 年，名为 Corporacion Antiguo 的公私合营机构为马德罗旧港的复原精心设计了一个令世人充满期待的总体规划。

1996 年，在与布宜诺斯艾利斯市政府签订协议后，Corporacion Antiguo 在全国范围内发起了竞标，中标方

案最终于 1999 年付诸施工。此次竞标只针对整个旧港区域内新建绿色空间的设计以及滨水地区的复原，因为旧港口的其他区域现在已经成为集商业与住宅为一体的综合地带。旧码头已得到修复，桥梁成为了联系各条人行道之间的纽带，不计其数的建筑物拔地而起。

设计团队由建筑师、城市规划师和景观设计师组成，他们面临的一个最大的概念性构想就是如何将一系列零散的城市空间整合成一个连贯的体系——需复原的滨水地区以及周边的白杨路，三条将旧港区与市中心联系起来的林阴大道，两个新建的大型公园。

所有这些组成部分都是分阶段进行施工的。最初完工的是林阴大道，与其同时完工的还有滨水地区的复原工程以及周边具有历史意义的地区；其次是

Micaela Bastidas 公园以及 Mujeres Argentinas 公园。在克服了各种难题之后，这个令人期待已久的新景观终于向世人展露其美丽的容颜。

展现历史风情的滨水地区复原

被当地人称之为"河畔南岸"的滨水地区建于 1918 年，设计的初衷是要建一条人行道，然而在马德罗港口区域被污染和废弃之后，这个地区逐渐被人们遗忘，到了 20 世纪 70 年代，这里几乎完全从人们的生活中消失。

设计师围绕人行道、灯柱和现有的围墙进行了一系列的复原工程，并对具有历史意义的白杨路上生长

状况较差的白杨树进行了替换。

邻近滨水地区的南端是一个三角形地块，设计师将这里重新规划为游乐区，并新建了两个盥洗室。几十年前被拆除的圆形剧场也得以重建，新建的剧场为具有半圆形屋顶的场所，周末人们可以在这里进行各种活动。

连接市中心的林阴大道

设计师新建了三条直线形步行道，与市中心新区直接相连，并将划分码头的桥梁有机整合起来。新建的林阴大道配有长椅、照明设施、自行车棚和垃圾箱等设施；双排树（每条林阴大道所栽种的树种各不相同，

1　入口广场一景
2　仙人掌
3　Micaela Bastidas 公园内绿树成荫

以展现植栽的季相变化。）与围绕坐椅的几何形灌木彰显出简单明快却又不失整体感的植被规划理念。

Micaela Bastidas 公园

Micaela Bastidas 公园建于 2000 年 ~ 2002 年，占地面积约 8 万平方米。公园的名字取自一名秘鲁女英雄，马德罗港新建的所有街道的名字都是以在拉丁美洲历史上赫赫有名的女英雄而命名的。

公园总体规划的特征在于多元化的空间。尽管公园在总体规划中占有较大的空间，而且被定义为开放的公共空间，但设计师认为它仍然可以被划分成紧密相连的几个小型空间。为了实现这一设计构想，设计师规划出三个面向西侧人行道的小型空间，从建筑方面突出空间的差异性，采用三座巨大的格宾（5m 高的建筑结构，内部填充本地岩石）划分出空间的界限，并分别以不同的设计元素加以修饰。根据岩石的形状和尺寸，用大小合适的石笼网将其笼罩在内。这种设计不仅突显了本地岩石的自然特征，还创造出一种柔性的曲线美。

公园北端的第一处空间是采用 2m×2m 的木板围

合成的绿色延伸带，游人既可以躺在上面休息，也可以将其作为野餐的餐桌；第二处空间是一个玫瑰花园，4000 多种玫瑰花呈几何形状种植，花圃中间设有狭窄的小径和坐椅；最后一处空间的设计具有更为突出的多样性，不仅栽种了种类繁多的植物，还设计了一个游乐场并安放了坐椅。

由于公园多变的地形，坡地的自然景观可以全面展现，人们可以从高处俯瞰整个公园以及周边的各种公共设施，人工铺就的斜坡和阶梯从草坪中穿过，可作为过渡来弥补从不同角度观看所带来的不适感。

在公园的最高处有一条蜿蜒的主道，两侧栽有当地乡土树种，如 tipu tree、合欢、美人树、蓝楹、珊瑚树。并在公园最高处栽种了各种颜色的开花灌木、草本植物以及当地草种，用以区分阶梯边缘并界定其他空间。天气晴好时，人们可以在草地上、长椅上沐浴阳光，或者在已经长成参天大树的树阴下席地而坐。

与滨水地区重新建立联系是这个项目另一值得称道之处，在周末的时候，很多人为了品味绿阴大道上的历史风情而特地从布宜诺斯艾利斯的南部地区赶到这里，这里还建有一个市场以及其他的娱乐设施。

Mujeres Argentinas 公园

Mujeres Argentinas 公园建于 2005 年～ 2007 年，占地面积近 10 万平方米。其名字源于一位阿根廷女性，这再次验证了马德罗港所有新建街道的命名规则。

与 Micaela Bastidas 公园不同，Mujeres Argentinas 公园具有明显的都市特征。公园的轮廓由一座巨大的格宾来界定，其轮廓和布局都非常大众化，突出了这

个城市的普遍规划原则。与 Micaela Bastidas 公园空间
差异性的特点不同，Mujeres Argentinas 公园完全是由
一个中心空间构成，朝向码头和其他一些较小的空间
开放。

格宾从公园的中央位置开始延伸，界定了一条连
接南北两端入口广场的主通道。主通道通往斜坡，南北
入口广场与西侧人行道处于同一水平面，格宾逐渐降低
高度与主通道平行。在这种设计理念下，主通道成为公
园最高处的一条环路和一处适合观景的较高位置。

中央区域是一片开放的波浪形草坪，被南北走向
排列的白杨树隔开，围绕草坪的东西向延伸出一条走
道，在走道的下面隐藏着公园排水系统的主管道。

设计师在公园东部的格宾后设计了一处比较随意
的空间，并设置了舒适的坐椅。为了营造出一片绚丽
的遮阴华盖，设计师在这里栽种了许多树种，大多数
为本地品种，长成大树后将非常醒目。这些树大多为

1 沿着狭窄的小径栽种的紫薇属植物
2 玫瑰花园中种植有 4000 多株不同种类的玫瑰花
3 在 Micaela Bastidas 公园前的新居住区周围开发
   的一处公共空间——这里栽种的本土植物犹如一
   片绿色波浪，为人们提供一处放松的场所
4 从码头古老石桥边缘卸下来的花岗岩石展现了原
   址的历史文化

珊瑚树、Bahuinias candicans、tipu tree 和梓树，这些树种在不同季节开花，看上去非常茂盛。

两处入口广场均为简约的城市风格：白色的地板、呈几何形状分布的混凝土长椅和旁边种植的菩提树；而格宾中的岩石和攀缘的蔓生植物以及白杨树形成的绿色屏风则在城市风格中渗透了自然风格。

设计师在邻近南广场的位置设计了一个与旧港口历史有关的参照物——将从沿码头而建的古老石桥卸下来的约0.3m高的方形花岗岩石呈几何形状排列，中间种植蒲苇，营造出一种非常和谐的自然景观。

在这个项目的植物配置中还有一个特别之处——那就是从阿根廷中北部移植而来的仙人掌。将这些仙人掌按种类排列，展示在步行道一端凹陷的方形区，

为此处的景观增添了一抹色彩。

Mujeres Argentinas 公园仍处于发展中，许多空间尚需时间来定义和完善。随着树木的成长和公园设施的战略性布局，这里必将成为另一个值得参观的名胜景观。如今，人们或者悠闲地在公园里骑自行车、滑冰、慢跑、散步，或者惬意地靠在斜坡上晒太阳，或者在广场上闲坐啜饮。

马德罗港新建绿色地带与滨水地区复原项目被誉为布宜诺斯艾利斯的城市规划锦上添花之笔，而这样一个充满活力、焕然一新的地区所发生的巨大变化更令这个城市的居民欣喜不已。这个项目使人们深信只要有全方位的规划设计以及充满创意的绿色空间设计，就有机会达成愿望，成为民众深爱的公共空间。

Buenos Aires, capital city of Argentina and one of the largest in the world, was born as a port-city and consequently grew and shaped in close relation with the river on whose shore was once founded.

The first port, called Puerto Madero, was finished in 1897 and less than twenty five years later a new port had to be built because the installations were already too insufficient. The old port's land was abandoned right away and the waterfront that had served as a public promenade and bathing area was abandoned some decades later. It was not until 1989, after the proposal of many revitalization projects that were never carried out, that a private-public partnership named Corporacion Antiguo Puerto Madero elaborated an ambitious Master Plan for the site's revival.

In 1996, after signing a contract with the City Government, they called for a national competition whose winning project was finally put into action in 1999. This competition only aimed at the design of new green spaces for the whole area, which now was a mix of commercial, residential and business uses, and the restoration of the

1 Mujeres Argentinas 公园内呈几何形状分布的长椅
2 林阴大道的一侧栽有本土树木
3 与格宾相邻的 Mujeres Argentinas 公园内主道，道路两旁栽种着柳树。（图片提供：Jimena Martignoni）

original waterfront. The old docks had been renovated, the bridges turned into pedestrian connections and countless new buildings were being planned.

The winning design team was formed by architects, urban planners and landscape design consultants who worked together for the entire definition of the project. What they determined to be raised as the most important conceptual premise was the interconnection of a series of urban pieces as part of one well-defined system: the restored historic waterfront and the poplar alleys adjacent to it, two new large parks and three pedestrian boulevards that would connect this area, at the same time, with the city centre.

All these pieces were built in phases. The boulevards were the first ones to be completed, together with the restoration works of the waterfront and the historic spaces around it; the first park called Micaela Bastidas was next and Mujeres Argentinas Park became the last one. Although the construction phases experienced many delays and were even put on hold several times and, in addition, the budget for the projects suffered considerable

restrictions and cutbacks, the new so long-expected image for this area of the city can be fully appreciated today. Born and developed in a quite unstable economy, this large project can be exhibited as a very successful one.

The Historic Waterfront

This waterfront, locally known as Costanera Sur (or Southern Riverside), was originally built in 1918 as a public promenade. With time, and especially after the abandonment of the first port's area and the slow process of contamination, this place started to fall into oblivion; in the 1970s Costanera Sur was already completely neglected.

The winning team proposed just a series of restorative works for the pavement. On the historic poplar alleys they just substituted dying trees and replanted others.

Contiguous to the waterfront, at the south portion, there is a triangular piece of land that was redesigned and refurnished as a playground area and provided with two new sanitary buildings or restrooms. And lastly, what had been the place for an amphitheater also demolished decades ago was renovated as a semi-roofed stage where people gather during the weekends, attend free shows or gym classes.

The Connecting Boulevards

These three linear pedestrian pieces were built completely anew and connect the new area of Puerto Madero with downtown. Aligned with the existing bridges that divide the historic docks, these boulevards were furnished with benches, lighting, bicycle racks and trash cans. A simple but consistent planting plan is based on a double row of trees (every boulevard with a different kind in order to guarantee diverse blooming seasons) and geometrical shrub arrangements that enclose the seating spaces.

Micaela Bastidas Park

Micaela Bastidas was built between 2000 and 2002 and

is approximately 8 hectares. The name of the park, which alludes to a Peruvian heroine, is part of the fact that all new streets in Puerto Madero were named after Latin American historically significant women.

What spatially defines this park is scale diversity. The designers determined that even when the park was naturally identified with the large scale of the Master Plan and the kinds of urban spaces it would serve, it should also generate some smaller areas related to a more intimate neighborhood scale. In order to achieve this, they laid out three different spaces that open towards the west sidewalk level and which are furnished with different elements; the architectural components that made this space differentiation possible were three rocky gabions that demarcate each one of the three areas. The gabions are some 5m-tall structures made of local rock. Put together according to shape and sizes into wire mesh containers, the granite rocks provide not only a remarkable naturalistic look to these gabions but the necessary flexibility to create soft curves.

The first one of the three spaces, on the northern area of the park, is a green extension dotted by some grouped 2m × 2m wooden decks on which visitors lie down or even use them as picnic tables; the second space is a rose garden where more than 4,000 species are geometrically exhibited and accompanied by narrow paths and seating spaces; the last one is a more diverse area where plant arrangements combine with a playground area and other seating spots.

Because the park has a dynamic topography it also presents different levels and natural slopes from where one can overlook the site and the urban surroundings; paved ramps and floating stairs interrupt the green of the lawn to negotiate these changes in the grading.

At the highest level of the park runs a zigzag-shaped main path which is completed with an interesting presence of native trees such as Tipuana tipu or tipu tree, Albizia julibrissin or silk tree, Jacaranda mnimosifoia, Chorizia speciosa or floss-silk tree and Erythrina cristagallis or coral

1 公园四周的新建筑物与田园风情的草坪形成鲜明对比
2 人们在 Mujeres Argentinas 公园的小径上漫步或者练习溜冰
3 滨水地区复原后的景象

tree. In addition, the highest areas of the park display an attractive colorful palette of flowering shrubs and herbaceous species and native grasses are used to mark the edges of the stairs and define other spaces.

During sunny days the site is crowded, either with people sunbathing on the grass, the park's benches or sun chairs they bring along, or with people seated under the nice shade that is provided by the many already grown trees planted here.

The reconnection with the waterfront area is another significant achievement, especially during the weekends when people from the south of Buenos Aires city (where green spaces are lacking) come to this area to explore the historic promenade. A food-market and fair is offered here as well, together with other recreational activities.

Mujeres Argentinas Park

Mujeres Argentinas was built between 2005 and 2007 and is almost 10 hectares. Its name translates as Argentine Women and, once again, refers to the naming of all new streets in Puerto Madero.

This park, in contrast to Micaela Bastidas, has a markedly metropolitan character. Its shape as much as its

layout in regards to the general and historic planning of the city, both want to express this clear vocation.

In this case, the silhouette of the park is defined by one single gabion. Instead of generating different spaces as in the first one, the site is structured with one central topographically dynamic space that open towards the docks and some smaller spaces that serve it.

As the gabion develops around the central space it also defines a main path which reaches two accessing squares respectively located at the north and south corners of the park. When going down to these two areas, whose levels coincide with that of the west sidewalk, the path turns into ramps and the gabion reduces its height to correspond with them. In this manner, the primary footpath is outlined as a loop which, at its central portion, is developed at the highest level of the park thus acting as an upper area from where to have general views of the site and surroundings.

The central area was thought of as a bare undulating lawn surface only interrupted by rows of poplar planted in the north-south direction of the grading and a second path that crosses it entirely from east to west. Underneath the path are hidden the primary pipes of the drainage system of the park.

Behind the gabion, on the east side of the park, there is a less formal space where one can find some more intimate seating spaces. In order to create a colorful dense canopy the designers planted a number of trees, mostly natives, whose tops will be visible from the park once they're fully grown. The species here are Erythrina falcata or coral tree, Bahuinias candicans or cow's foot, Tipuanas tipu or tipu tree and Catalpa bignoniodes or Indian bean; thought to provide diverse blooming seasons throughout the year, these trees already look luxuriant.

The two squares have an almost austere urban look with their white paved floors, the geometrically disposed concrete benches and linden trees planted on a grid pattern. However, there is a handsome sense of nature here given by the sight of the rocky gabions and the vines already climbing up them and the green screens of poplars that appear from behind.

Adjacent to the south square, the design team incorporated a very interesting reference to the old port's history: the granite pieces that were removed from the rims of the old stone bridges along the docks. These whitish 11 in-tall square-shaped rocks were set in groups and form geometrical arrangements on the ground; rows of pampas

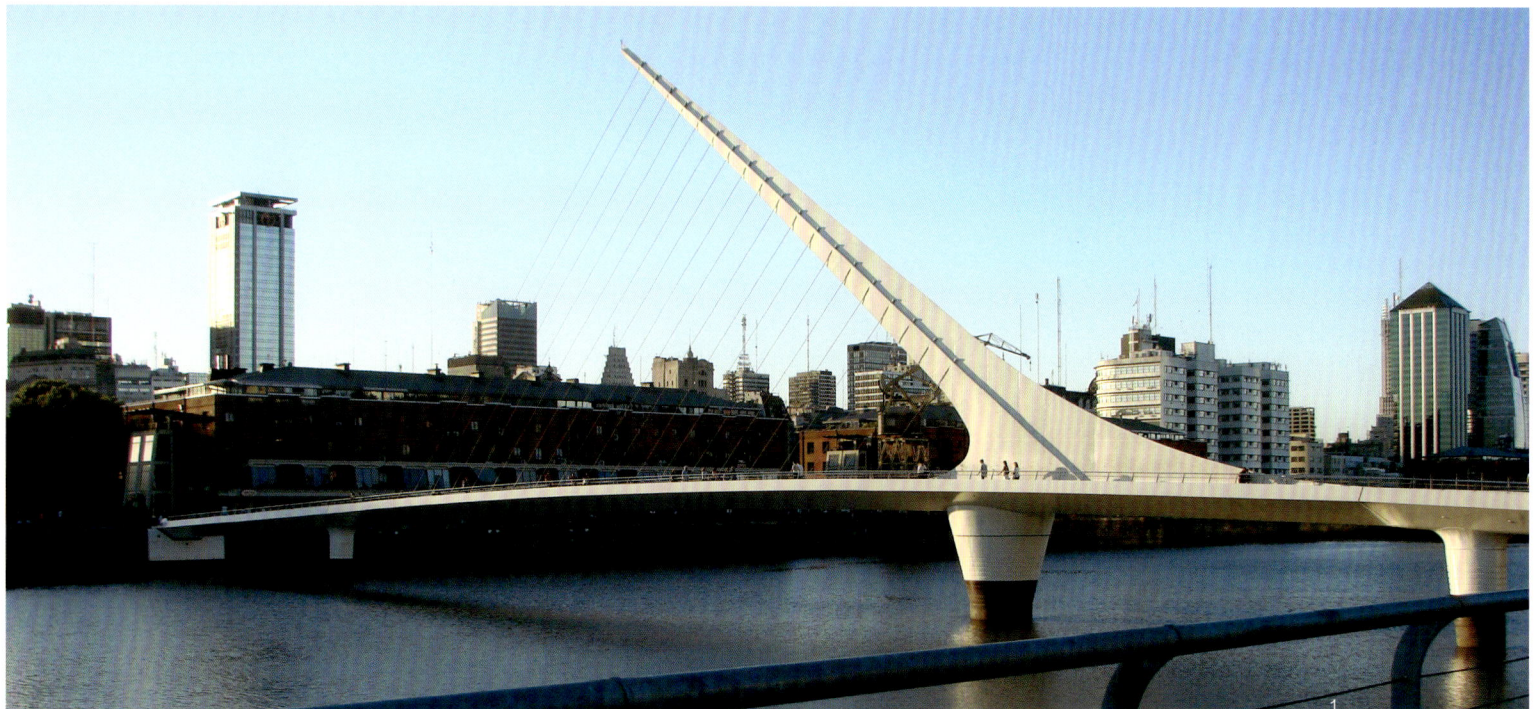

grasses (Cortaderia selloiana sp) are planted between the groups of rocks and provide a very naturalistic image to the site.

Completing the planting plan, there is a particular incorporation of cacti species specially brought from northern and central areas of Argentina. These cacti are exhibited in two sunken square-shaped areas located right at the end of the central path, in front of one of the sidewalks. The plants are grouped by species and add a colorful and sculptural component to the place.

This park is still in very early stages and some of its spaces need some time to look completely defined and inviting. But with many of the trees already growing copiously, the furniture strategically placed and the paths and ramps offering a dynamic experience for visitors the site promises to be another interesting place to visit. Now, people ride their bikes, practice roller-blading, jog and walk the paths of the park, others just lie down on the slopes or sit around at the squares while drinking mate or just sunbathing.

The project for the new green areas of Puerto Madero and the restoration of the waterfront as a whole has proven to be a successful addition to the city of Buenos Aires, especially in a brand new area whose growth and final image have been perceived by its people as a quite dramatic change. The most obvious (and probably familiar) lesson here is that when supported by well-thought out, innovative and joyous green spaces, new developments have a better chance to fulfill what at least should be their most desired purpose: to become beloved places.

狭长线性公共空间的扩展 —— 西哈莱姆码头公园

Expanding the Long and Linear Open Space — West Harlem Piers Park

撰文：Barbara Wilks   图片提供：Johannes Feder Alison Cartwright   翻译：李沐菲

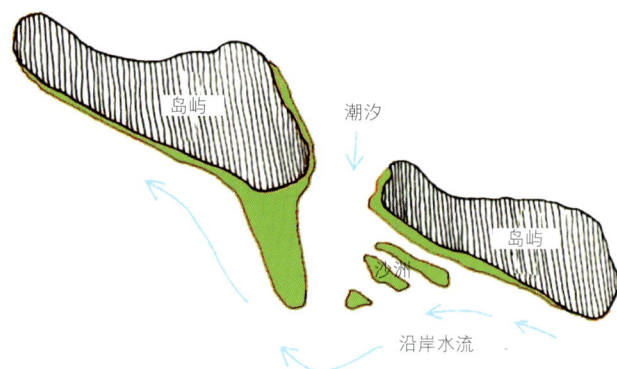

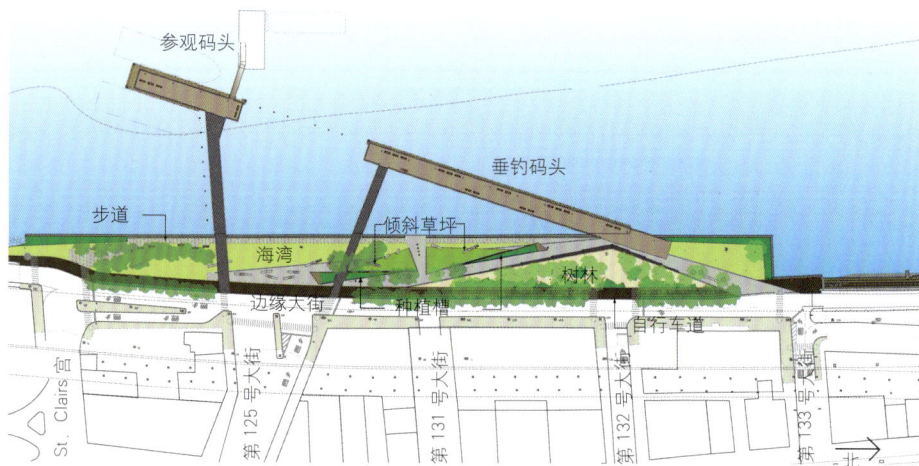

参观码头
垂钓码头
步道
倾斜草坪
海湾
树林
边缘大街
种植槽
自行车道
St. Clairs宫
第 125 号大街
第 131 号大街
第 132 号大街
第 133 号大街
北

该项目位于一片狭长的线性区域，这一公共空间的诞生使哈莱姆社区与哈得孙河得以重新连接在一起。该地区原本是一座面积仅为 6410 ㎡ 的停车场，现已扩建为占地面积达 9804 ㎡ 的公园，以一种极富意味的方式重新诠释了这座城市与河流间的界限。该项目也是该社区 30 年来为重获滨水地区景观而不断努力的结果，与河流的联系使整个社区都焕然一新。

该公园位于哈得孙河沿岸的第 129 号与第 133 号大街之间，宽度还不及一座网球场，与其相邻的街道以及高速公路将这一区域与周围的社区分隔开来。从前，这里是峭壁之间的一个天然的小海湾，直到近些年才成为一个工业港口，后来曾被改建为带有护栏的停车场。该项目的主旨是在哈得孙河沿岸将曼哈顿林阴大道延伸。除此之外，设计目标还包括与客户及社区进行通力合作，重新使该社区的居民与滨水景观建立联系。在社区成员积极的支持下，设计团队举行了一系列的公开会议，共同促成了该地区总体规划的诞生。规划的第一阶段便是西哈莱姆码头公园的建设。

设计策略

　　为了使此处成为一个实用的聚会地点，不仅需要一些线性的连接，而且公园的面积也亟待扩大。为了增加公园的占地面积，相关人员采取了多项策略。首先，通过兴建新码头来增加公园的实际面积，这是社区规划中的一个关键部分。码头一直以来与街道网络相邻，并延伸到水面，而设计师所采用的新策略则是通过河口沙洲的形式使码头呈一定的角度，与其相连接的匝道在水面上形成了一个环形的通道，巧妙地扩展了公园的空间；另一项策略是通过缩窄相邻公路、减少不必要的路面宽度来增加公园的面积，街道路面的重新规划使公园面积增加了25%。最后，该公园在视觉空间上的拓展则是通过将码头拓展到步道中，从而增加对角线的长度来实现的，塑造出更加深长的远景效果。同时这条对角线也界定了该公园的两个主要区域：即开放海湾和树林。

　　开放海湾是哈莱姆主要街道——第125号大街的尽头，引导游人来到公园，也是最主要的开放性的聚会地点。在最初的设计中，设计师在该区域设置了码头，但是社区居民坚持要求能够从第125号大街直接看到开放的滨水景观。最终的设计方案接纳了这一要求，码头在海湾处被中断，取而代之的是与街道走向相同的匝道直通水面。通往这里的其余街道或中断、或形成与街区结构一致的通路，也形成了一些特殊的空间，如位于第131号大街上的宝石般璀璨的水景；而其他诸如长椅、种植槽以及草坪等景观元素的设置，则如同退去的潮水般随意自然。

　　树林从第132号大街引至公园的入口，并从街道开始沿一系列台阶逐级下降，一直延伸到接近水面的地方。这些台阶既可以供人们小坐休息，又能为欣赏水景的人们提供一些荫蔽。林阴地中的其他坐位则是一些看似随意布置、实则是在施工中精心摆放的未经加工的花岗岩石块。树林里还有一些由艺术家Nari Ward设计的护栏，上面有社区的题词，唤起了人们对该场地历史的回忆。

设计要素

为了能够适应广泛的公众群体，该项目的规划需要具有多元化的特点。垂钓是该社区的一项传统活动，因此一座形状规则且整洁的垂钓平台是必不可少的。公园北端的皮划艇区适合开展一些水上的娱乐活动，也是出入河流的通道；公园的最南端则主要是为通勤的人们提供的"水上的士"，在将来也可能提供轮渡服务。在两座倾斜的"沙丘"之间设计了一些自动喷泉，孩子们可以在炎热的夏季来此戏水。社区居民的另一需求便是能够在水畔进行一些节奏稍舒缓的活动，因此又设计了一些其他项目，如位于公园东部边缘的曼哈顿自行车道以及河边的公共步道。

项目建材

公园所使用的建材体现了场地的历史及河流景观的特点——海湾区的大型三角形种植槽与该场地的海运历史相呼应；从废弃物中清理出的混凝土块在此也得到了再利用，形成排水洼地；另外，路面所使用的石块是哈得孙河中的鹅卵石和经过筛选的当地石材。

可持续性

社区居民本着强烈的主人翁责任感积极参与到公园的设计过程中，而这种参与和投入也成为了西哈莱姆码头公园成功的关键因素。合作设计中为提升可持续性所采取的措施如下：

· 1672 ㎡的自行车道被改建成为公共步道，不仅提升了社区的生活品质也减少了交通拥堵；

· 将有铺装的高地的37%改建为透水表面，既有利于雨水的迅速排放，又有利于提升空气质量以及改善热岛效应；

· 由废弃混凝土块铺就的排水洼地能够将雨水及时导向既定的排水口；

· 曼哈顿自行车道的延伸为骑自行车上下班的人士提供了进出哈莱姆社区的安全路线；

· "水上的士"服务为上班族提供了另一种节能的交通方式；

· 本地建材（花岗岩石块、哈得孙河的鹅卵石以及碎石）的使用；

· 为减轻码头建设对环境的影响，河边设置了"圆礁"，可供水生动物栖息；

· 均选用耐盐性良好、符合纽约市树木栽植标准的本土植被，不选用易受病虫害的本土植被，如红枫。

The West Harlem Piers Park transforms a long and linear site into a public space that reconnects the Harlem community to the Hudson River. A narrow 69,000 sf parking lot has been expanded into a 105,526 sf park that re-imagines the threshold between city and river in a meaningful way. The West Harlem Piers Park is the culmination of a 30 year struggle by the community to regain their waterfront and renew their neighborhood through connection to the river.

The site, no wider than a tennis court between 129th and 133rd Streets along the Hudson River, had been cut off from the neighborhood by the adjacent streets and highway above. This point of access to the river, historically a natural cove between adjacent bluffs, and more recently an industrial port, had become a paved and fenced parking lot. The project intent was to create a continuation of the Manhattan linear Greenway along the river. Additionally, our goals included working with the client and the community to reconnect the residents to the water. Working closely with a strong group of community activists, the design team held a series of public meetings that resulted in a master plan for the area. The first phase of the plan was the creation of West Harlem Piers Park.

Design Strategies

In order to create an effective gathering place, in addition to the linear connections required, the space of the park needed to expand. Several strategies were employed to increase the size of the space allocated for the park. The first was to literally expand it through the construction of new piers which were a key part of the community's vision. Historically piers aligned with the street grid, continuing out into the water. Looking for a new strategy, we used the model of sand bar formation to angle the piers. Connecting gangways created a loop of circulation out over the water, expanding the space. Another strategy was to expand the space by narrowing the adjacent roadway, eliminating unnecessary lane width. The reconfiguration of the street provided a 25% increase in park area. Finally, the park space was expanded visually through the extension of the diagonals created by the pier configuration into pathways that create long vistas. This diagonal structure also defines the two main areas of the park, an open cove and the shaded woodland.

The cove is the terminus to 125th Street, the "main street" of Harlem. It welcomes visitors into the park and is the main open gathering space. In earlier designs we had placed the piers in this area, but the community was firm in its vision of seeing open water down the 125th Street corridor. Incorporating this desire into the final design, the pier is broken at the cove and the connecting gangways follow the lines of the street out into the water. Other streets that terminate at the site create breaks and pathways that align with the street grid and create special places, like the jumping jewels water feature at 131st Street. Other elements in the cove like the benches, planters and raised lawns are arranged as if deposited by the receding tides.

The woodland provides an entrance to the park from 132nd St. The woodland rises from the street, and then cascades down towards the water in continuous steps. These stairs provide seating and function as a shady overlook to the water. Other seating in the woodlands is created from scattered remnants of the original granite bulkhead which were replaced during construction. Pieces of the guardrail with inscriptions by the community created by artist Nari Ward are also arranged among the undergrowth, recalling past memories on this site.

Program Elements

A diverse program was necessary for the park to cater to its wide array of user groups. For community members a fishing pier provides a proper cutting and cleaning station

that honors the long standing tradition of fishing in this area. A kayak float at the north end of the site encourages water related recreation and allows access to and from the river. For commuters, the transportation pier at the southernmost tip of the site will soon offer water taxi service and is configured for future ferry service as well. For children, self activated fountains located between the two sloping "dunes" allows for water play in hot weather. For active recreation, the Manhattan bike trail follows the eastern edge of the site, honoring the community's desire for slow paced passive recreation to be located near the water like the required 12' waterfront shore public walkway which edges the water.

Design Materials

The materials used in the park reflect the history of the site and the river environment. The large triangular corten planters in the cove area echo the marine history of the site and concrete cobbles scavenged in demolition were reused to create drainage swales. Additional materials include exposed aggregate paving created from Hudson River pebbles and stone screenings made of local stone.

Sustainability

The community has a strong feeling of ownership of the park, due in part to their involvement in the design process. Overtime, this involvement will develop into stewardship and is crucial to West Harlem Piers Park's success. Measures that were implemented in the collaborative design and that promote sustainability are:

• The return of 18,000sf from the vehicular to the pedestrian realm enhances community life and decrease traffic congestion.

• The replacement of the paved upland with 37% pervious surfaces will positively impact storm water runoff, air quality and heat island effect.

• Swales created from reused cobbles channel storm water to existing outfalls.

• The extension of the Manhattan bikeway provides safe access in and out of Harlem for commuting bicyclists.

• Water taxi service will offer other energy efficient forms of commuting.

• The materials (granite blocks from the site, Hudson River pebbles, and stone screenings) are all locally sourced.

• As mitigation for the pier construction, "reef domes" have been placed in the river to provide marine habitat.

• Plant material is salt tolerant, native in part, and meets NYC standards for tree planting, avoiding native species which are now prone to pests and disease like the red maple.

主防线 —— 荷兰新水系

Main Defence Line — New Dutch Waterline

撰文 / 图片提供：OKRA Landscape Architects　　翻译：王玲

隐藏在优美景色中的主防线时断时续、时隐时现，仿佛披上了一层神秘的面纱。

主防线是"荷兰新水系"的一部分，在施工过程中充分利用了原有的堤坝和自然下沉区。荷兰水系分为三个区域，分别是湖区（Vecht区）、乌得勒支区和河区，每个区域都有自己独特的景致和防御系统。主防线包括密集广阔的防护区，形成一个多元素组成的复杂系统。沿防线分布着诸多的要塞、水闸和其他小型水利工程以及后来修建的炮台。大量穿越堤坝的沟渠尽管不能直接看到（这些沟渠的位置只能在设计图上看到），却是溢洪体系的重要组成部分。

景观创造机会

主防线贯穿三种不同景观——它在每种景观中的表现各具千秋，景观亦为它提供了丰富的想像空间。湖区的水面美轮美奂，仿佛即将溢出一样；乌得勒支区随处可见的胆小鬼雕塑栩栩如生、惟妙惟肖。主防线与城市肌理完美地融合在一起，河区的景致大多藏于侧堤相对连续的一侧。人们在这里不仅能够最真切地感受到主防线的延绵，而且只有在这里才有可能看清主防线和道路、要塞、炮台的关系。

干预

全长85km的主防线无论是自然风光还是其休闲性都颇具声望。因为大量的沟渠和邻近基础设施的存在，正在修建的自行车道尽可能地沿着主防线的轨迹进行。主防线在功能上应该是连续性的，而整体性取决于自行车道的总体印象。

设计中充满诗意的一面是将"沟渠作为景观干预的标志"，巧妙地展示出该项目的细节设计，使得部分被隐藏的防线一目了然。

两处精心之作

· De Gagel 要塞

De Gagel 要塞守护的通道是诸多需要沟渠来加快溢洪过程的通道之一。一条狭长的泻水台仿佛薄纱一般沿堤坝展开，精细的切边两侧是错落的水台。因为泻水台是不对称地穿过通道，人们的视线常常被引向南面——沟渠在一个平台上被继续延伸到更开阔的地方。主防线与原来的一条道路相连，路的尽头是一个观景台，人们可以在这里看到战争时散布在战场上的掩体。一条蜿蜒小路引领着人们走回要塞，旁边的一条小路则通向 Ruigenhoek 要塞。

Honswijk 要塞中隐藏的社区道路连接了莱克河（Lek）北部的许多防御工事，也是荷兰新水系的组成部分。原来隐藏的水面、道路和堤坝，在荷兰新水系的规划中都被设置在该项目中的休闲场所的视线范围内。除了修建和翻新建筑物、土岸和观景台之外，自

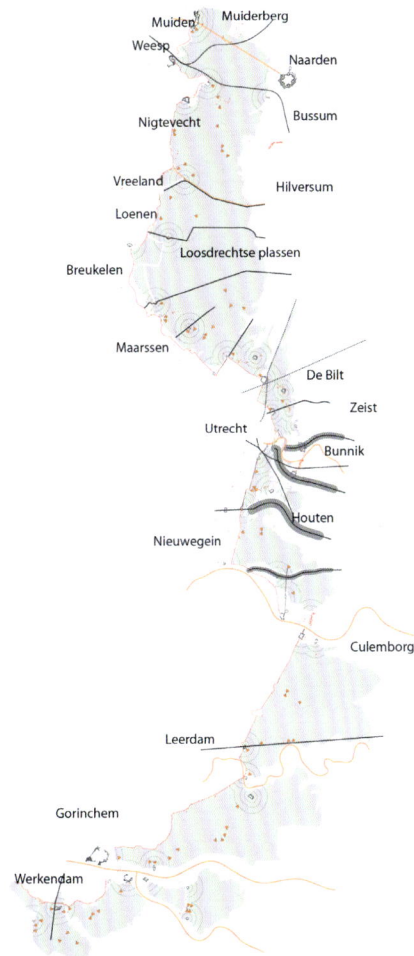

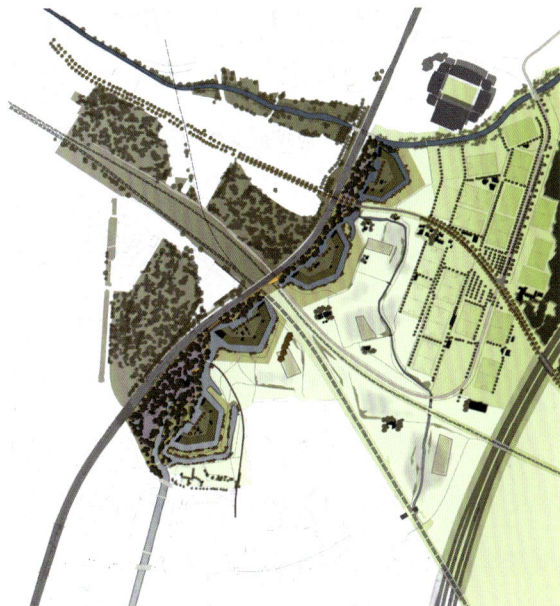

主防线是荷兰新水系的一部分

行车道和步道的修缮也增强了该项目的休闲功能。

　　Honswijk 要塞中隐蔽的社区道路并不是防水堤而是一个防御要塞，而作为系统的主防线其不仅仅是一条具体的线路，更需要精细性和明确性。沿着堤坝的通道不仅能够让人们深刻地体会到它的防御作用，还为人们提供休闲散步的场所。通过地面通道的一个入口人们可以抵达水面。南面的一系列阶石形成了越过堤坝的楼梯，使防御工事更加有效。

　　主防线的规划是在干预的基础上打造一处富有趣味性的景观，并通过这些景观使防线显得独特而不突兀。

· Tienhoven 要塞

　　Tienhoven 要塞周边的景观有着悠久的历史，目前的空间设置为人们展现了立法的发展历程。随着时代的变迁，景观的用途不断变化，其景色也随之改变。在该项目的规划中，沿线的景观设计已经按照未来的发展变化被制定出来。

　　Tienhoven 要塞建于 1848 年至 1850 年，由一个 15m×15m 的防御警卫室组成，旨在关闭 Tienhovensche 运河，保卫附近的码头、水坝水闸和 Kraaiennester 水闸。

乌德勒支——眼镜堡（Houtense Vlakte）

　　始建于 1822 年的眼镜堡是保卫乌得勒支的首道防御线。1843 年，人们在这里修建了通往阿纳姆的火车道。四个眼镜堡成为荷兰新水系的又一亮点。这种独特的防御工事在沿线的其他地方都不曾出现。眼镜堡前方的区域很难被淹没，因此有必要在这里修建更多的防御工事。此外，这里两个重要的通道——Kromme Rijn 和 Koningsweg 也需要防御。

　　荷兰新水系的整体性受到了国内民众的广泛关注，而且它也是荷兰结构蓝图和绿色结构等市政规划的重要组成部分。

Hidden and tucked away in the landscape the main defence line forms a secret line. There has never been talk of a continuous lining. The main resistance line in a way is just an unclear and diffuse line. The main defence line is a special line; hidden and tucked away in the landscape it forms a secret line.

The line is part of the "New Dutch waterline". At the construction the existing embankments and natural depressed areas are used as much as possible. Therefore the Dutch Waterline can be divided into three zones: the lake area or the Vecht area, the Collar of Utrecht and the river area. Each characterized by its own landscape and defence system. The line contains intensive and extensively defended zones. In its totality it forms a complex system of elements. Spreading along the line are a sequel of citadels, an enormous amount of smaller objects such as sluices and other water-scientific works and the later added casemates. Not directly visible, but an essential part of the inundation system, are a large number of cuts to be made in the embankments.

Landscapes offer opportunities

The main resistance line runs through three different landscapes and behaves different in each one. The Dutch Waterline complies with the underlying landscape. The three landscapes offer different opportunities for the imagination of the Main defence line. The elaborate water in the Vecht area and the lake area offer an opportunity to see the waterline in a state of inundation. The opportunity around Utrecht is the large number of recreants who are able to reach the Main defence line in a short period of time. The Main defence line is intertwined n in the urban tissue. The opportunity for the river area is hidden in the relatively continuous lines over the side dikes. The feeling of a continuous Main defence line is strongest there. Especially in the river area it's possible to extricate the relationship between Main defence line and intensifying in the form of access, citadels and casemates.

Interventions

For the full 85 kilometre the line gets a report of scenic and recreational articulation. The ongoing bicycle route will follow the trace of the main resistance line as much as possible. In a functional way it is about creating a continuous line, which is important, because of the large number of cuts and parallel adjoining infrastructure. A lot depends on the first impression of the route.

The poetic side of our approach is creating a "cut as a sign of intervention". By making cuts, particularisation can be connected to explanation. The means of a cut is suitable, to make (partially hidden) systems visible.

Elaboration of two special places
• The Gageldijk

The access defended by fort De Gagel was one of many that were pointed out with a cut to fasten the inundation process. A long and narrow water-table is made through the dike, accommodating a thin membrane of water. With an excellent cutting edge on both sides, an increased water level is represented. Because the table asymmetrically cuts the access, the view of the person passing is directed to the Southern directions, where the line of the cut is continued in a platform directed to the openness. The line connects to an existing path and ends with a viewpoint from which one has a beautiful view of the scattered series of bunkers in the field. A winding path leads the person back along fort De Gagel; a side path takes them to fort Ruigenhoek.

The covered community road at Fort Honswijk connects a number of defence works at the North-side of the Lek, as part of the New Dutch Waterline. The water, the roads and the embankments have a hidden purpose, which is put back on stage within the scope of the recreational development of the New Dutch Waterline. Besides reconstruction of the buildings, re-profiling the earthen bank and making a viewpoint, partial recreational use can be enhanced by improving cycle and foot paths.

The Covered Community Road at Fort Honswijk is not a damming embankment, but a defence embankment. The main defence line as a system calls for particularization and explanation, much more then concrete lining. By adding the "cut" through the embankment, the defensive character of the embankment has been made experiencable and a short walk is made possible. The water lying behind becomes accessible by an entrance through the cut in the ground. Southwards, a series of step stones form a staircase over the embankment, making the defence work accessible.

With that, the plan for the main defence line is based on making intervention producing a readable landscape, by which the line does become special but not always and everywhere.

•Fort Tienhoven

The landscape around Fort Tienhoven has a rich history. The current spatial construction is the result of this legislative history. The use of the landscape has gradually changed over time, whereby the image of the landscape changes. For the New Dutch Waterline, the line perspective has been drawn up, which adapts to the future changes in the line landscape.

Fort Tienhoven was built in 1848-1850 and consisted of a defensible guardhouse of 15 x 15 meters. Its purpose was the closure of the Tienhovensche Vaart and to guard the adjoining inundation quay, a dam sluice and the Kraaiennester sluice.

Utrecht—Lunetten

The Lunetten are part of the first belt of defence-works around Utrecht, the construction was started in 1822. In 1843, the railway to Arnhem was added to that. The four Lunetten form a special part of the New Dutch Waterline; you will find this form of defence nowhere else in the line. Since the area in front of the Lunetten (the Houtense Vlakte) was difficult to inundate, it was necessary to add more fortifications. Besides there were two important accesses here (the Kromme Rijn and the Koningsweg) that needed to be defended.

Further, there is a lot of national attention for the New Dutch Waterline as completeness and the NDW is also mentioned as an important aspect in the municipal plans such as the Structural Vision and the Green Structure plan.

漫步于湾前海岸 —— 澳大利亚布莱顿海滩公用步道

Biking along the Bayside Foreshore — Brighton Beach Shared Path, Australia

撰文：Courtesy Tract　Emporis 网络编辑　Pedro F Marcelino　　　图片提供：Courtesy Tract　GoogleEarth　　　翻译：刘建明

　　南澳大利亚布莱顿住宅区位于维多利亚州墨尔本市郊东南方向约 12 千米处。布莱顿在湾前市辖区内，大部分住宅区依菲利浦港湾而建，形成一处城市海滨地带。滨海公路是一条非常受自行车运动爱好者喜爱的路线，海滩上还设有越野自行车赛道，南北走向的湾前越野自行车赛道穿过布莱顿的海滩。这一区域内设有多处休闲空间和娱乐设施，其中"路北小栈"就是一个深受游客喜爱的咖啡屋，在这里可以欣赏到菲利浦港湾的壮观景象，还可以眺望墨尔本市的城市轮廓线。布莱顿的主海滩是历史悠久的"布莱顿浴场"的所在地，这里聚集了墨尔本市大部分的富人以及豪华住宅区。

　　湾前市政府计划在海湾街和布莱顿中部浴场之间修建步行道和自行车道，以加强二者之间日渐松散的联系，并将公共开放空间、娱乐设施与拱形的"沿海公用步道"项目整合起来。

　　在与湾前市政府以及该项目的海滨区域设计顾问进行磋商后，设计师提出了一个动态的解决方案：架高的蜿蜒的木栈道可以使游客自如地出入海岸边的各种景观设施，尽情地在沙滩上和海面上游览。具有保护作用的近海堤坝最大化地保证了沙滩不受海水侵蚀，大面积的沙滩为游客聚餐提供了场所。其中的一道堤坝邻近布莱顿皇家快艇俱乐部，以促进码头附近的细沙沉积。道路顾问的工作还包括植被再生和维护、

沙滩维护、道路周边和木栈道以及邻近街区的绿化工作等。

　　架高的木栈道为步行和骑自行车的游客提供了一种独特的行走体验——游客可以凌驾于海面和沙滩之上，穿行于平滑的曲线之中，感受着与传统陆地旅行截然不同的行走方式。

　　一系列详细的景观规划理念和图片集锦为设计提供了便利，也让周围的居民能更清楚地理解设计意图和方式，便于项目的建造。

　　设计师将该项目的设计看做是对普通道路设计的一种尝试，采用一种新颖的设计方式。景观设计师与规划师在设计一处线性场地时产生了全新的想法，通

项目总规划图

过对石墙、堤坝和高架平台的整合，为港湾道路系统的缺失部分提供了一个完美的解决方案。该项目所产生的效果非常理想，最直接的效应就是提升了布莱顿海滩的整体形象。布莱顿海滩公用步道的景观设计方案超越了设计预期的效果，为使用者提供了丰富的海滩活动场所，而不仅是一条海滨道路。所有的使用者(行人、自行车运动爱好者和轮滑运动爱好者)现在都可以随意使用海岸上的各种观景设施——驻足观赏或是在观景码头欣赏海岸的美景，也可以在东面的沉积沙滩上玩耍。这里还设有宽阔的场地供人们与宠物玩耍，设计师并将这块区域设在邻近的沙滩上，用围栏将其围合，尽量不影响到其他使用者。

JAN 2005   MAR 2005   JUNE 2005   AUG 2005

ROCK WALL

新筑堤坝的延伸部分

布莱顿中部码头

新栽植的灌木丛

新筑的堤坝

沙滩区

布莱顿步道

露天甲板

Sandown 大街入口

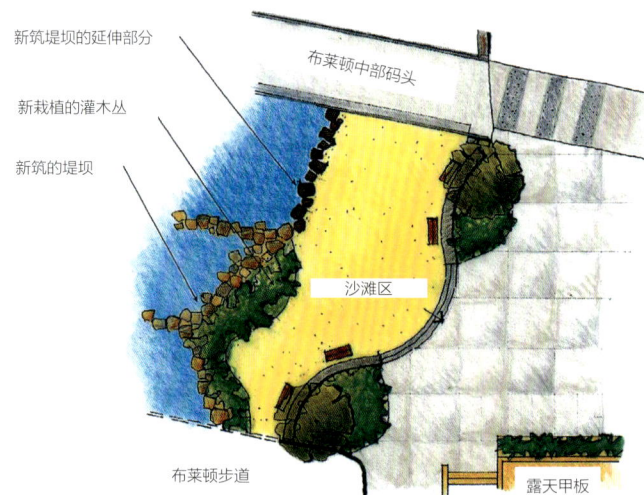

Sandown 大街入口和堤坝

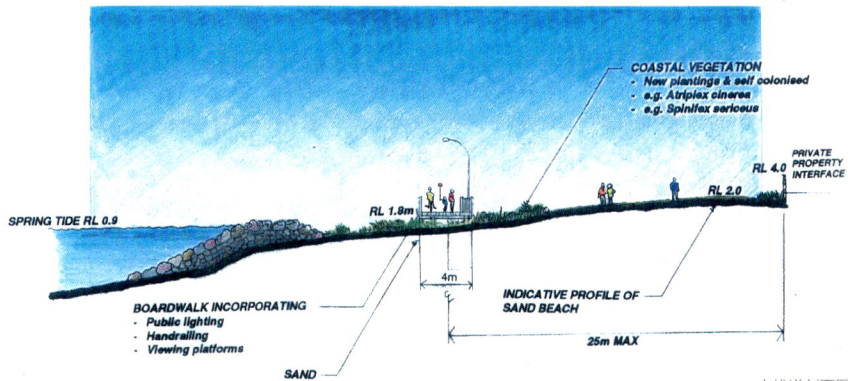

木栈道剖面图 1

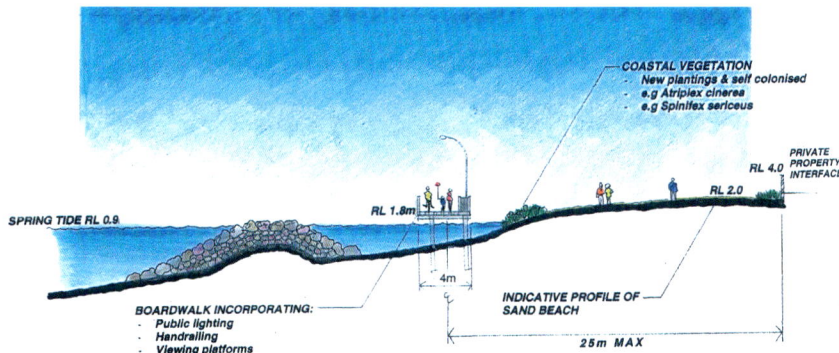

木栈道剖面图 2

Named after its English namesake in 1840, the residential area of Brighton in southern Australia is a suburb of Melbourne, State of Victoria. It is located approximately 12 km to the southeast of the city. Brighton is included in the Bayside Council, and much of it lies on the shores of Port Phillip Bay, making it somewhat of a beachfront urban area. Beach Road is a popular cycling route, while off-road cycling tracks are also available at the beach. The Bayside Trail off-road bicycle path goes through the Brighton foreshore, continuing north and south. On this area, several civic structures and recreational facilities are available, among which the North Road Pavillion, a highly popular cafe with wide views of Port Phillip Bay. The beachfront also affords stunning views of the Melbourne city skyline at the distance. The main beach in Brighton is also home to the historic Brighton Baths. This suburb is reputed for being the home to some of Melbourne's wealthiest people, with matching housing estates.

Tract Consultants were engaged by Bayside City Council to project a well-balanced landscape design solution for the missing pedestrian/bicycle link between Bay Street and the Middle Brighton Baths, integrating the public open space and recreational facilities of the area in the overarching Shared Coastal Pathways Project.

1   海岸边的暗礁、木栈道和观景平台

2   木栈道上的观景平台

3   步道下方在海中显现的巨型暗礁

Working with the City Council and the project's marine consultant GHD, Tract generated a dynamic solution consisting of a curvilinear, elevated boardwalk that enables visitors to connect with the coastal setting, traveling over sand and water. The integration of a series of protective and offshore rock groynes maximizes the foreshore by encouraging the re-nourishment of the beach's sand strips, thus offering opportunities for picnicking and other bay front social activities. One of these rock groynes lies close to the Royal Brighton Yacht Club, encouraging sand buildup near the jetty (refer to plan). Tract's work included the re-vegetation and fencing, maintenance of the sand spit and beautification works along the adjacent pathway and eastern path access points works, interface areas between the boardwalk and the adjoining streets.

The elevated boardwalk offers a unique experience for pedestrians and cyclists as they travel over-water and sand, in a sequence of smooth curves, contrasts with the traditional on-road path as experienced north and south of the intervention area.

The generation of a series of detailed landscape concept plans and photomontages assisted in the process of community consultation and enabled the community to understand the design intent and approach, making Tract's work easier.

Tract regarded this project as an opportunity to attempt a unique approach to a fairly common on-grade path link. By integrating rock walling, groynes and an elevated platform, the project's landscape architects and planners were able to generate a fresh design response to a relatively straightforward project brief of providing a missing link within the existing Bay Trail. The results are enriching and enhance Brighton's beach side character. The landscape solution for the Brighton Beach Shared Path exceeded the design brief by offering users a rich coastal experience, rather than a simple path link. Users (pedestrians, pets, bikers, roller-bladers) can now properly engage with the coastal setting, pause to take in the extensive views to the west via a series of viewing decks protected from the main path, or play in the sand (in the re-nourished beach to the east). There are off-leash recreation areas for users with dogs, on the adjacent sand spit, which has been strategically fenced to minimize conflict with other path users.

1　人们在布莱顿海滩的公用步道上悠闲地骑着自行车

2　木栈道也是宠物及其主人的休闲场所

3、4　在公用步道上悠闲散步的人们

5　在布莱顿海滩公用步道骑自行车

6　蜿蜒的木栈道

Malecon 2000 滨水空间

Malecon 2000 Waterfront

撰文：Douglas Dreher　　图片提供：Carlos Rodríguez S.　　翻译：刘建明

休闲区

　　Malecon 2000 项目中的休闲区南起市民广场（P. Icaza 大街），北接弗朗西斯科—德奥雷利亚纳大街，占地面积 11 195.32m²，在老码头的旧址上兴建而起，设计师将该项目扩展至河面，发展成为一个综合型娱乐休闲中心。

　　游客可以从南部的市民广场或北部的弗朗西斯科—德奥雷利亚纳大街的入口进入休闲区，两处入口都设有保安、监视器和公共汽车候车亭。Junín 大街上还有一个机动车和行人的入口，可以通向即将建成的林阴大道或停车场的入口。绿化带和现有的植物有专门的维护流程，并新增加了一些植物品种。

　　这里的公共照明设施和其他区域一样，也分为三种：一种是为步行道提供足够的灯光照明；另一种是为突出景点、纪念碑、楼梯和标志等的灯光照明；最后一种是为了增强码头和河边景色的灯光照明。此外，Simon Bolivar 大道的周边采用金属围栏加以围护，给人以安全感。

　　步行道作为南北方向的主要通道，设计师将其设计为沿着河流的走向延伸，而通用服务网络设施（如电气、饮水、排污、加油和电话等）位于步行道的下方。

　　位于步行道和 Simon Bolivar 大道之间的公共空间

从南至北依次为：

·货运广场

在这个广场上，游客可以看到厄瓜多尔主要的铁路设施，并设有专门的展览区，人们可以通过顶部带有凉棚的休息平台进入展区。广场周围设有休息区和乘凉区，还有一座与周围环境十分协调的雕塑，寓意其与铁路的密切联系。

·儿童游乐场

儿童游乐场紧邻货运广场，有一条形状类似雪橇底部的斜坡通向瞭望塔。这里有适合各年龄段儿童玩耍的游乐设施，也有专为游客准备的坐椅和绿色长廊休息区，还有一个溜冰场。

·喷泉广场

喷泉广场位于 Junín 大街的中轴线上，两侧设有餐厅和盥洗室。喷泉广场可作为机动车进入停车场的入口，中心是一个重建的铁铸喷泉，周围被彩色的灯光环绕着。

·餐厅和盥洗室

在 Junín 广场的两侧、临近儿童游乐场的位置有两座建筑物：一座建筑物的一楼是盥洗室，二楼设有一个自助餐厅和一处可以俯瞰河流的瞭望台，人们可以在这里交谈、观景和休息；另一座建筑物设有餐厅，

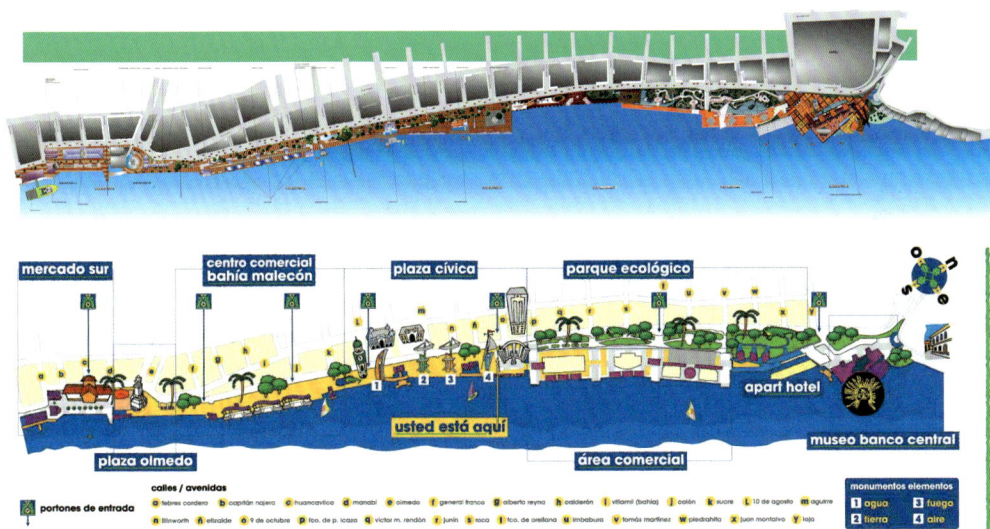

在其周边是专门用做室外运动和有氧运动的空间。

·运动区

运动区位于 Junín 和奥雷利亚纳大街之间的绿化带，配备了有氧运动和户外运动设备。

·奥雷利亚纳广场

位于奥雷利亚纳大街的入口处，也是位于休闲区北端的一个公共空间，作为连接 Malecon 花园的纽带。在广场的中央部分有一个椭圆形空间，主要用于保护一棵古树。

花园

花园紧临南部休闲区，北接 Loja 大街，占地面积

为 35 650m²。花园共有两个入口，分别为南部的奥雷利亚纳入口和北部的 Loja 入口，这两个入口处均设有保安、监视器、公共汽车候车亭以及出租车停靠点。

花园的布局与市民广场以及休闲区南北走向的布局一致，河边的散步道与河流的走势平行；位置较高的散步道下面是一个能够容纳 320 辆机动车的停车场；散步道和 Simon Bolivar 大道之间是占地面积约 22 000m² 的 Malecon 花园，栽种了厄瓜多尔当地的植物。总长约 1.5 千米的散步道将现有的植栽有机地联系起来，此外还栽种了一些新的植物，以实现最初对植物园区的设计理念，而面积广阔的绿色区域更丰富了植物园区的内涵。

花园的观景区和休息区中的散步道穿行于不同的公共空间，如广场、小广场、瞭望台、露台、堤坝、绿色长廊等。这些公共空间则通过溪流、泻湖、岛屿、喷泉、小桥、斜坡和路径等景观元素来点缀，可以在此举行一些活动，如植物展览会和纪念品展销会。

所有的人行道都要确保残疾人、老年人和童车能够顺畅通行；区域内设有保安和监视器以及方向指示标识和图文说明；长椅、垃圾筒、电话等设施均分布于此；通用服务网络位于奥雷利亚纳和 Tomas Martinez 大街之间，如电气、电话、排污、雨水排放和机械化灌溉等设施十分齐全，从这里直至 Loja 大街，分布在公园和停车场之间。

位于 Tomas Martinez 大街和 Loja 大街之间的停车场刚刚完工，这里还是"捐赠纪念碑"的入口，"捐赠纪念碑"上铭刻了 50 000 多个在 Malecon 2000 项目中做出过贡献的个人或机构的名字。北端的一座建筑物主要包括天文馆和海洋博物馆。

Recreational Area

This sector is located following the Civic Square (P. Icaza Street) in the south up to Francisco de Orellana Street in the north, and it has an extension of 11, 195.32 m^2, developed on the surface of the old pier, foreseeing for the future an expansion on the river with an entertainment center.

People can access here from the Civic Square in the south and through a gate on Francisco de Orellana Street in the north, both entrances are provided with control, surveillance, information and bus shelters. On Junín Street there is also a vehicular and pedestrian entrance that will lead to the future boulevards or entrance to the parking areas of the commercial and entertainment blocks. Special care has been given to the green areas, the existent trees have been subjected to a process of care and new species have been added.

As in the other sectors, the public illumination is of three types: one that provides enough general illumination for the pedestrians, another that accentuates the points of interest, monuments, trees, stairways, signs, etc. and one that enhances the border of the pier and the River. Also, the whole perimeter toward Simon Bolivar Ave is walled by a metallic fence that provides it with the corresponding safety.

Being the main circulation south-north-south, a pedestrian walkway that runs parallel and close to the river has been designed. General service nets (electricity, drinking water, sewage, gas, telephones) will be developed under the mentioned pedestrian walkway, having conceived the necessary connections for the future constructions.

Between this pedestrian walkway and Simon Bolivar Ave the following public spaces are developed from south to north:

•Wagon Square

In this square we can find an Ecuadorian railroad adapted as space for exhibitions to which people can access through a waiting platform with a pergola. The square is supplemented with resting and shading areas, there is also a sculpture that integrates itself and is part of the wagon environment and which has relationship with the railroads.

•Children Playground

This area, following the wagon square towards the north, begins with a ramp that leads to a watchtower with a toboggan. It is equipped with games for children of all ages; areas with pergolas with seats and shade are available for the visitors. A skating ring accessed independently is a complement to the playground area.

•Fountain Square

It is located in the Junín street axis, flanked by two constructions that house spaces for cafeterias, food and restrooms for the public. This square will serve as future vehicular entrance to the parking spaces of the commercial areas that will be developed over the river. In its center, the presence of a restored cast iron fountain and with color lights following its original image stands out.

•Buildings with Cafeteria and Public Restrooms

Two constructions flank Junín Square, based on shipping metaphors and adjacent to the children playground, one of them houses on its ground floor the necessary restrooms to cover the demand of the external areas; its upper floor is a covered terrace with a cafeteria and river mirador. The access is through some stairs which, as at the Civic Square, serve as meeting, observation and rest places. The other twin construction houses a restaurant bar with terrace, and it is adjacent to the areas dedicated to outdoor exercises and aerobics.

•Exercise Area

Developed between Junín and Orellana streets and it consists of a green area equipped for aerobics and outdoors exercises.

•Orellana Square

Located in front of Orellana street gate, this is the northern end of this sector and it serves as connection with the Malecon Gardens. In its central part an elliptic space has been designed to house an old existent tree, complementing it with a water fountain and a Japanese garden.

Gardens

The Gardens are developed right after the Recreational Area in the south, up to Loja Street in the north, on an area of 35,650 m². The access to the Gardens is through

the Orellana Gate in the south and Loja Gate in the north, both entrances are provided with control, surveillance, bus shelters and taxi stops.

Just as in the Civic Square and Recreational Area the circulation is mainly south-north-south, thus a high pedestrian walkway, which runs parallel and close to the river, has been designed; under this walkway there is a parking lot for 320 vehicles; between the walkway and Simon Bolivar Ave lies "The Malecon Gardens", a great park of approximately 22,000 m^2 of extension, where according to didactic approaches, the different vegetable species of the Ecuador are located. Additionally it has an artificial stream on Orellana Street and a pond that spreads between Tomas Martinez and Loja streets. It has a 1.5 km-long pedestrian path ensemble, and this circuit has been adapted to the presence of the big existent trees, having had special care in supplementing them with new plants so that the botanical zoning proposal is achieved; this zoning is enhanced by extensive and profuse green areas.

The walk in the Gardens, includes view and rest areas supported by public spaces such as squares, small squares, miradors, gazebos, bleachers, jetties, pergolas. These spaces are supplemented with streams, lagoon, island, fountains, bridges, ramps and paths where activities related to the park such as scheduled exhibitions and sales of plants and souvenirs from The Malecon Gardens are developed.

All the walkways guarantee the circulation of handicapped, elderly people and baby carriages giving the whole community access to the park. The walkways have arranged hierarchically by dimension and by finishing

material. The whole area is provided with control and surveillance, as well as general orientation and didactic signaling, these aspects are supplemented by the PA system that allows the control, announcement of events and in turn to reinforce sounds that recreate the ecological botanical sector atmosphere. The urban furniture is prepared strategically and it consists of benches, drinking fountains, garbage bins, telephones, etc. General service nets (electricity, telephones, drinking water, sewage, rains water drainage and mechanical watering) have also been foreseen for the correct operation of the park.

Between Tomas Martinez and Loja streets, next to the river a 320-car park has been developed. Its roof serves as pedestrian platform for the view of the river and the gardens, and as access to the " Monument to the Donors" that symbolizes and perpetuates the collaboration of more than 50,000 natural and juridical people that contributed economically to make a reality a good part of the Malecon 2000. In the north end, there is a building dedicated to a Planetarium and the Maritime Museum.

海边起伏沙丘

Coastal Rolling Dunes

撰文：Peter McGuckin SLR Consulting　　　图片提供：James Brener SLR Consulting　　　翻译：王玲

总规划图

2004 年，SLR 咨询公司首次被委任负责 Blyth Links 地区的整体规划。在社区和市民的积极参与下，一项全面且富有想像力的规划被制定出来，对创建这个能够吸引居民和游客、令人激动的新公共空间起到了指导作用。

该项目涉及 Blyth Links 改造区，这片荒芜的公共空间是人造沙丘的一部分，邻近 Blyth 炮台，该炮台是 1916 年专为海防而修建的。沙丘在岁月的洗礼中不断地被侵蚀，使得该地区逐渐丧失了能够激发和保留公共兴趣的特色或基本的基础设施。

该项目的设计概要在许多方面都极富挑战性，尤其在两个方面需要达到平衡。首先，绵延的海岸线上需要更多全新的防风浪建筑，而海滩上多数景观的魅力却源于大海所带来的宽阔无垠的感觉——几乎没有高耸的建筑物打断一望无际的海景。其次，虽然设计希望能留住老游客并吸引新游客，但沙丘是一个相对脆弱的环境，增加的人流量会对沙丘造成更多的侵蚀。因此，设计需要将人流引向特定的区域——最好是沙丘上的自然沟壑带，或者是可加固和围合的地块。

　　总体规划方案也充分考虑到一些其他因素，如与附近新开发的住宅相连接、为战时设施和救生设施提供适当的环境和有效的交通动线。设计改善了一些正规的游乐设施和停车场设施，关注预期增加的游客兴趣。

　　实施方案的焦点是新中心广场的建设。通过一个新的多功能开发项目将原来的步道（日后将进行翻新）和海滩与 Blyth Links 的西部区域连接在一起。新广场的亮点是一系列巨大的木质防波堤，与放置在海滩上防止沙子流失的对应设施相似，但是它们却与这里寒冷的东北风向反向设置，不仅充当了挡风设施，而且还是一种艺术媒介。各种海洋艺术品被嵌入防波堤上透明的树脂块里，防波堤上被雕刻出波浪和水泡的形状，照明系统也被嵌入其中。粗犷朴实的防波堤有效地防止了人为的蓄意破坏，同时其阶梯式的轮廓确保了沙丘和海景的开阔视野。

1、2　防波堤上的艺术品
3　Blyth Links 地区的中心广场

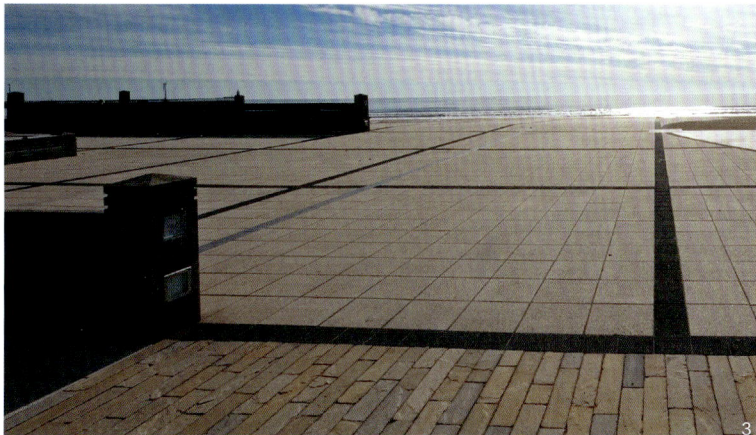

　　鲜明的防波堤主题在圆形露天广场和围合的下沉式游乐场中继续得以展现，彰显出材料与其运用的自然手法，以及地形地貌的充分利用，特别是木质游乐设施和安全沙地的广泛使用。环形木板围栏将空间有效地围合起来，宠物狗被禁止入内。游乐场根据儿童的年龄划分出不同区域，设置了更多适合幼儿及其家长活动的游乐设施，木质防波堤则将这些设施保护起来。

　　海岸线上广泛分布的休闲设施包括交通静音限速措施和减少侵蚀、穿越沙丘的便捷的人行道。设计也对战时设施——英国保存最为完好的海防炮台和观察哨进行了生动的解读；印有文字的道路铺装、标识和街边景观小品也成为设计中的新艺术品。

　　该项目的总体规划使得 Blyth Links 地区成为游客和当地居民向往的理想场所。无论是被动性娱乐还是主动性娱乐，个人水上运动和冲浪运动在此备受欢迎。2008 年初，一座独特的服务楼在这里修建而成，包括公厕、淋浴、水上运动爱好者的更衣室、急救点以及销售热饮料和小吃的摊位。楼内设有海滩救生员食宿区，其中包括一座观察海岸情况的瞭望塔。在当地俱乐部的建议下，服务楼内还备有独木舟、水上运动设备和救生设备。该服务楼成为 Blyth Links 地区的焦点，并且鼓励通过提供高品质的设施来实现特色建筑长期的可持续利用。

1　木质防波堤围成的圆
　　形露天广场
2　游乐场
3　中心广场 1
4　Blyth 节上游玩的人
5　圆形露天广场

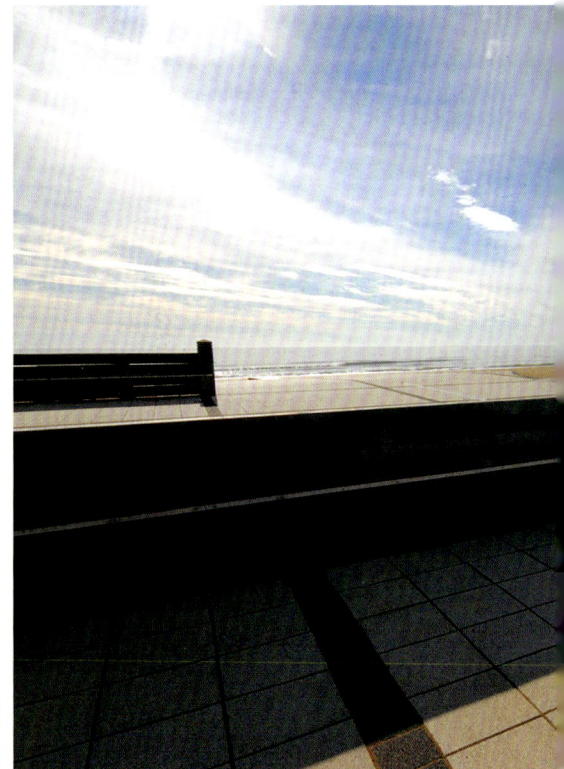

SLR Consulting was first appointed in 2004 to develop a comprehensive and imaginative masterplan for the Blyth Links area. Following significant community and member engagement a comprehensive plan was prepared to guide the creation of an exciting new public realm to help attract residents and visitors.

This project concerns the Links Regeneration Area, a section of reclaimed dunes, the principal feature of which is an area of desolate public realm adjacent to the Blyth Battery—an artillery emplacement built for coastal defence in 1916. The dunes were eroding and the area generally lacked any features or basic infrastructure to inspire and retain public interest.

The brief has been challenging in a number of areas but two aspects in particular required a balance between opposites. First, this extremely exposed stretch of coast called for provision of new shelter structures and yet much of the attractiveness of this stretch of beach derives from its feeling of openness—an expansive seascape with few tall structures to interrupt views. Second, although there was a determination to help retain existing visitors and attract new ones, at the same time the dunes are a fragile environment

and increased footfall could give rise to increased erosion. The solution would need to channel people to specific areas—ideally to those where natural gaps in the dunes occurred and the land could be consolidated and shelter provided.

The masterplan proposals also took account of a number of other factors including links to nearby recent residential development and providing an appropriate setting and access for the war time structures and life guard facilities. A need was also identified for improved formal play facilities and, anticipating increased visitor interest, improved parking.

The implemented scheme's major focus is a new central plaza connecting the existing promenade (planned for later refurbishment) and the beach with a new mixed use development to the west of the Links. A strong feature of the new plaza is a series of large timber groynes, similar to their counterparts placed on the foreshore to prevent loss of sand, but here placed counter to the prevailing cold north easterlies where they act both as popular shelters and as an artistic medium. Various maritime artefacts have been embedded in resin blocks set within the groynes, wave forms and bubbles are cut out of the structures and lighting inset. Their robust

nature has also proved resistant to opportunistic vandalism and their stepped profile manages to retain the feeling of open views across the structures and out to the dunes and sea beyond.

The strong groynes motif continues into an amphitheatre and a sheltered sunken play area stressing a natural approach to materials and play, making use of landform, predominantly timber play structures and sand as a safety surface. A timber ring fence effectively encloses the entire structure and keeps dogs out. The space is also zoned for different ages of children with more informal play facilities for toddlers and parents sheltered by the timber groynes.

Wider leisure improvements to the coastline include new car parks with traffic calming measures and easy-access pedestrian links through the dunes to help reduce erosion. Interpretation of wartime structures is provided—one of the best preserved coastal batteries and observation posts in the UK, and also new artworks included inscribed paving, signage and street furniture.

The Links' appeal to both visitors and local residents has been greatly enhanced through delivery of the masterplan. The area has become a popular destination for both passive

and active recreation with the use of personal watercraft and surfing becomes particularly popular. During early 2008 a distinctive amenity building was completed which features public conveniences, showers, changing accommodation for watersports enthusiasts, a first aid point meeting/utility space and a small concession for the sale of hot drinks and snacks. The building incorporates new accommodation for the beach lifeguards including an observation tower commanding unrivalled views of the coastline. There are storage facilities for canoes and other watersports/lifesaving equipment designed through consultation with local clubs. The amenity building provides a focal point for the Links and encourages long term sustained use of the area through the provision of high quality facilities in a distinctive iconic building.

1　在中心广场游玩的人们
2　防波堤
3　中心广场 2
4　Blyth 海滩

哈利戴维森博物馆

Harley-Davidson Museum

撰文：David Motzenbecker（oslund.and.assoc.）　图片提供：Michael Mingo　翻译：刘建明

1 哈利戴维森博物馆、酒店、广场

该项目位于密尔沃基市第六大道与运河街交叉处，占地面积为 80 937 ㎡。很多年前，这里曾经是一个制盐工厂，正因如此使该项目用地的土壤受到了污染。设计师通过场地分析和坡度划分，再结合古老的防波堤结构以弥补技术局限，最终决定将该项目场地地表覆盖 1.2m 厚的植入土壤。选用牧场非禾本草本植物和本土草本植物来再现梅诺莫尼河谷的滨水植物景观。依托新的土壤结构栽种的新植被能够最大限度地还原该项目用地的原有生态环境。

2008 年 7 月，占地面积为 12 077 ㎡、造价为 7500 万美元的博物馆项目正式对外开放，整个项目包括展会空间、饭店、咖啡厅、零售店、会议场所、特别活动设施及公司档案室等。博物馆的设计意在吸引国内外的游客前来体验哈利戴维森（Harley-Davidson）的人文、产品、文化和历史。博物馆记载了哈利戴维森公司详尽的发展历程和始创者的壮志豪情，以及员工、经销商、供应商、公司领导层与协会会员的故事和其他信息。

该项目设计流程中有一个鲜明的特色，即整个规划场地在设计之初就被视为"博物馆"，而不仅仅是一座综合性建筑。正是基于客户与设计师之间所达成的这种共识，博物馆的体验才延伸至游客所不能及的户外空间，这样游客们可以一边共享精彩的历史故事，

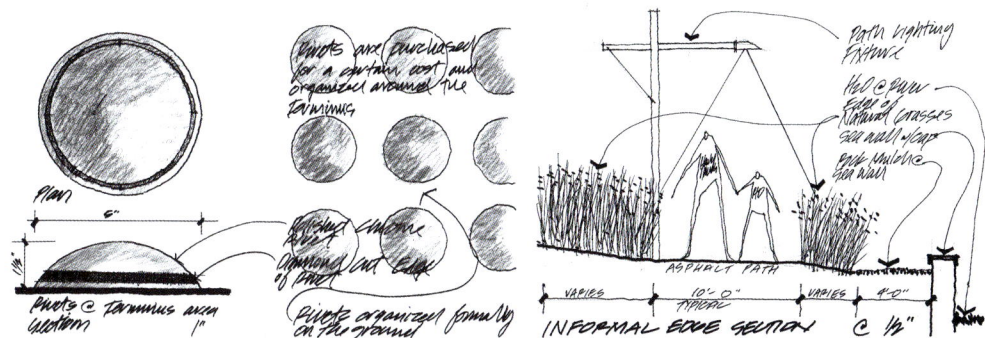

河流与河岸步道手绘图（图片提供：Tom Oslund）

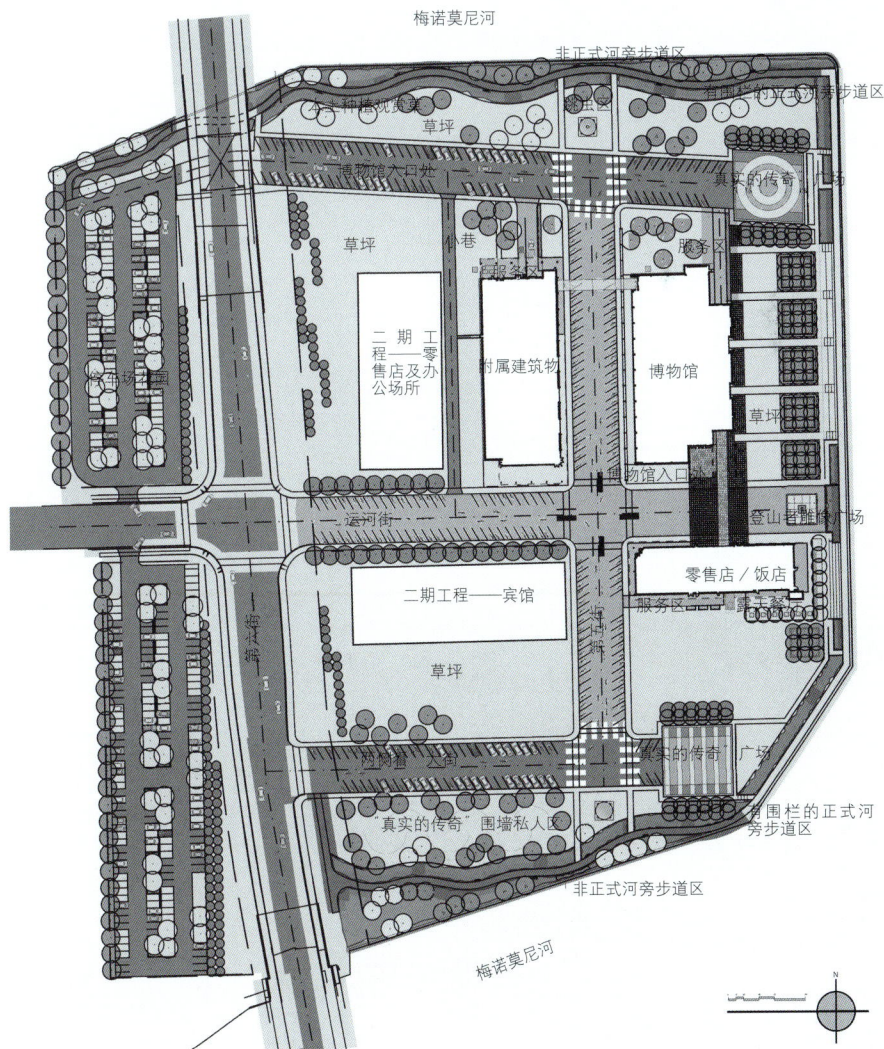

总平面图

一边欣赏梅诺莫尼河岸边的绿色空间。

该项目的规划融入了引人注目的城市设计元素，并与周边的水体和绿色空间完美而和谐地搭配。北面不规则的开阔草坪和南面的停车场将河畔步行道与规整的博物馆户外空间连接起来。停车场的部分车位采用嵌草铺装的形式，方便雨水径流及灌溉。奠基平台附近的河岸边缘沿线种植了成排的树丛，它们与博物馆构建了一处清凉的滨水空间。该项目为这处被遗弃了的工业化旧址搭建起了绿色的背景屏幕，这将有助于整合城市中心与梅诺莫尼河谷的景观。从极具特色的铆钉到广泛应用的钢结构可以看出，建筑物以及该项目所选取的材料反映了这一区域及哈利戴维森的工业化传统。

设计师在工艺衔接和本土化定制设计中发挥了重要的作用，例如定制工字型钢梁长椅、照明设备、种植槽、防波堤沿线栏杆和极具特色的铆钉。铆钉的设计构思具有双重意义：既可作为募捐用的装置——上面写有"真实的传奇"，又是哈利支持者们记载历史的

一种方式；同时设计师通过这种方式将哈利戴维森的支持者与他们所在的地域联系起来。铆钉是制造机动车的关键部件，在哈利戴维森的许多服装设计中也比较常见。铆钉可被抽象地认为是构成博物馆的重要组成部分——既是搭建博物馆骨架的工具，也是与博物馆整体联系紧密的媒介。游客可以购买铆钉，并在上面刻字，然后将其放置在两处可选位置的任意一处。较大的铆钉可安置在两处大型界标广场的地表；较小的铆钉可安装在弯曲的考登钢墙上。

　　界标广场位于东西向街道的两端——一处为方形，另一处为圆形。这两处广场专为"真实的传奇"这款铆钉所设计——一处广场主题为"经销商和哈利业主集团（H.O.G.）"，另一处主题为"哈利戴维森友人"。主界标广场上有一座青铜雕塑，描述了哈利戴维森发展史上标志性的时刻——爬坡赛。重新改造后的工业化旧址——从曾经用于生产颗粒状工业材料的工厂，到现今焕然一新的哈利戴维森博物馆——成为了该项目场地的核心，也是连接南北路线的枢纽。

　　该项目成为了可持续利用的典范，同时也是一个纪念和创造历史的地标，是密尔沃基市民在未来数十年里可以尽情享受的新兴绿色空间。

1 被草丛掩映的博物馆
2 被当地植被环绕的博物馆
3 主干道的左侧为商店和酒店，右侧为博物馆

1 河滨步道一侧的植物
2 自行车爱好者和行人喜爱的河滨步道
3 主界标广场上的青铜雕塑
4 秋日里的河滨步道
5 "真实的传奇"铆钉

The Harley-Davidson Museum is located on a 20-acre parcel of land at the corner of Sixth and Canal Streets near downtown Milwaukee. Many years ago a salt manufacturing plant once occupied the site, which resulted in contamination of the site soil. Through oslund.and.assoc. site design and grading strategies, combined with the technical fill limit imposed by the old seawall structures, it was decided that the existing site would be topped with four feet of import soil. This new soil will bring opportunities for restoring the site with new vegetation. Plant materials were selected to reintroduce riparian landscape of Menomonee River Valley which includes prairie forbs and grasses native to the region.

Opened in July, 2008, the 130,000 square foot, $75 million Museum development features exhibit space as well as a restaurant, café, retail shop, meeting space, special events facilities and the Company's Archives. The design vision is that the Museum will draw visitors locally and from throughout the world to experience the people, products, culture and history of Harley-Davidson. The Harley-Davidson Museum celebrates the rich history of the Company, the passion of the riders, the stories of the employees, dealers, suppliers, company leaders and community members and much more.

One of the unique elements in this process was that from the beginning the entire site was considered as the "Museum", not just the building complex. Because of this understanding between the client and the designers, the museum experience will extend to outdoor space where visitors from near and far can meet, share stories and enjoy the green spaces adjacent to the edge of the Menomonee River. These varied and fascinating stories are woven into the fabric of Harley-Davidson, shaping the legend and setting the stage for a bright future.

The plan for the Harley-Davidson Museum and its site incorporates striking urban design elements and engages the surrounding water and green spaces. A series of riverwalks, planted with native species, connect formal outdoor spaces of the Museum complex with informal open lawn spaces

and the parking gardens to the north and south. The parking gardens utilize grass pavers in some of the parking stalls to help with stormwater runoff and infiltration. Bosques of trees line the river's edge directly adjacent to the museum and the Founder's Terrace, creating a cool space next to the water to relax after a ride. The museum site brings a breath of green back to a derelict industrial site—this space will help to unite the city center with the Menomonee Valley. The building and site materials selected for the project reflect the industrial heritage of the area and of Harley-Davidson, as seen in the extensive use of steel members and rivets as visible features.

oslund.and.assoc. played a crucial role in crafting engaging, customized site vernacular elements by the design of custom I-beam benches, lighting, planters, the railing that edges the seawall and the unique memorial rivets. The rivets were conceived as both a fundraising mechanism—entitled "Living The Legend"—and as a way for Harley enthusiasts to become a part of history. "Living The Legend" was designed as a way to connect Harley-Davidson enthusiasts to the place. The rivet is key part in the construction of a motorcycle and also shows up on many Harley-Davidson clothing designs. The rivet is abstracted as a site component, designed to create a texture within the site, while providing individual connections to the museum. Individuals can purchase a rivet, have it engraved, and then have it placed in either one of two locations. Larger rivets can be mounted into the ground in two large terminus plazas on site; smaller rivets will be mounted to curving Cor-ten steel walls that create an industrial maze garden on the grounds.

The terminus plazas are located at the ends of the on-site, east-west streets—one is square in shape, the other circular. These two plazas have been created for the "Living The Legend" rivets—one plaza allocated to Dealers and H.O.G. (Harley Owners Group) Chapters, while the other is for the Friends of Harley-Davidson. The main terminus plaza holds a bronze sculpture depicting one of the defining moments in Harley-Davidson history—the Hill Climb. Reclaimed industrial hoppers—once used for granular industrial materials, and now painted Harley-Davidson orange—act as vertical focal points at either end of the on-site, north-south circulation route.

This museum is an achievement of sustainable reuse, a place for the recollection and creation of history, as well as a revitalized green space for the city of Milwaukee for generations to come.

Reso

度假区／酒店

Gudbrandsjuvet 峡谷

Gudbrandsjuvet Gorge

撰文 / 图片提供：Jensen & Skodvin Arkitektkontor AS

Gudbrandsjuvet 峡谷约 20 米深，位于著名的旅游景点 Trollstigen 和 Valldal 之间。关于这里有一个凄美的传说——16 世纪时，一个名为 Gudbrand 的男子带着自己的新娘从这座峡谷跳下，Gudbrandsjuvet 便由此得名。

瀑布

如今，Gudbrandsjuvet 因其壮观的瀑布、深邃的峡谷、陡峭的悬崖而成为一个远近闻名的旅游景点。每年春天，冰层融化的水都流入此处，吸引了大量的游客来欣赏这里壮丽的美景，停留在前往 Trollstigen 的途中。设计师建造了一系列的观景设施，如观景台、小桥、护栏、景观酒店等，表达出独特的设计语言。

观景台

观景台是由 25mm 的激光切割而成的钢质薄板构成，形成悬臂桥，将悬崖两端连接起来。栏杆的设计满足了各处不同的安全系数要求，向内部弯曲的曲线设计使游客可以放心地凭栏眺望壮观的瀑布。悬臂桥的材质因地制宜，不同部分各不相同，这样的设计十分合理并能够节省成本。停车场的观景台由预制混凝土构成，犹如一条自行车链，将各个角落连接起来。服务中心的设计也运用了相应的几何图形。

栅栏桥

这座桥位于奔腾的流水之上，公路桥之下，将游客引至峡谷裂隙处。整座桥由金属制成，主要结构是一系列不锈钢的金属链，吊在周围的岩石墙上，使桥变得十分轻便，同时也使人们更接近水面。

景观酒店

每个房间都是独立的，并且有一到两个墙面是由玻璃制成的，从房间内观赏到的室外景观壮丽独特，美不胜收。每个房间的地理位置都不同，因此不同的房间可从不同的观赏角度饱览室外四季各异的美景。

Located between the famous Trollstigen and Valldal, the Gudbrandsjuvet Canyon is about 20m deep. There is a sad story about it, which said that a man called Gudbrand jumped down into the canyon together with his bride in 16th century. That's why people called it Gudbrandsjuvet Canyon.

Waterfall

This site has become famous for its spectacular waterfalls. Though not very high, the place nevertheless fascinates because of the dramatic deep and narrow cuts in the cliff, solely a result of the river's continuous friction with the cliffs. Every spring huge amounts of melting-water from the glaciers creates a thundering roar at the site. The place gets several hundred thousand visitors every year, stopping on their way from the Geiranger fjord to Trollstigen. At the lay-by we have designed several smaller interventions; viewing platforms, bridges, a service center, and also a landscape hotel. The different interventions have related but independent architectural expressions.

Viewing Platforms

The main platform is constructed by 25mm laser cut steel sheets, cantilevered like a bridge around the cliff, hung in each end. The railing has a geometry that allows it to be continuous even with very different security requirements from place to place. The large inward curve allows the tourists to securely lean out over the deadly waters. The bridges are made from different materials according to what is most appropriate at each site. The platform at the parking side is made from prefabricated elements of concrete, like a bicycle chain, an element that is connected in the corners but rotated in the angle that will fit the site. This was appropriate at this site because cantilevering prefabricated elements had obvious advantages economically and practically. A related geometric concept is used for the service center.

Stocking Bridge

This bridge will lead the visitors deep into the crevasse, just above the roaring water, and under the existing road bridge. The structure is based upon the fact that basically only wires are needed to construct the entire bridge, except for the rings that gives the "stocking its volume". The whole construction is based on a net of thin stainless steel wire, hung up on the surrounding rock walls, making the bridge extremely light, and giving a very intense contact to the water below, both visually and otherwise.

Landscape Hotel

Basically each room is a detached small independent house with one, or sometimes two of the walls constructed in glass. The landscape in which these rooms are placed is by most people considered spectacularly beautiful and varied and the topography allows a layout where no room looks at another. In this way every room gets its own surprising view of a dramatic piece of landscape, always changing with the weather and the time of the day and the season.

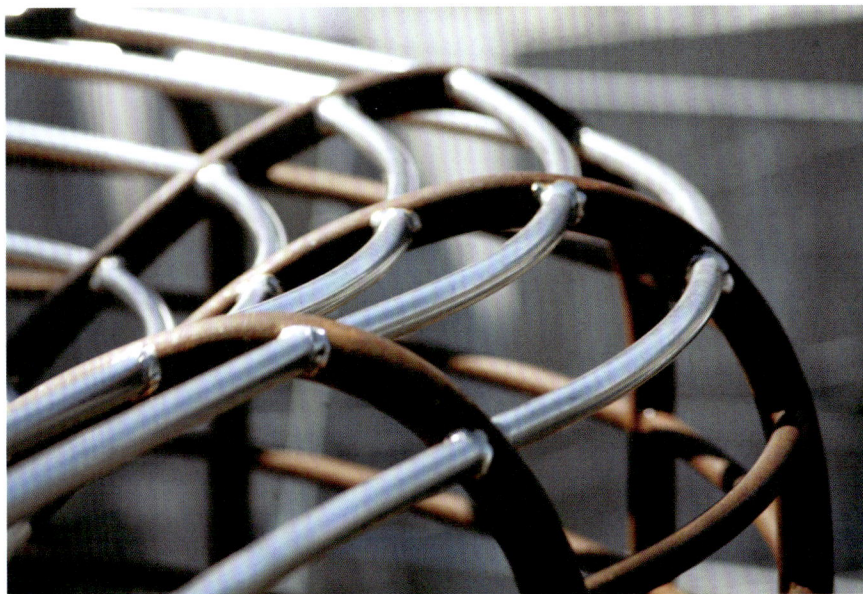

全新的马真塔海滨度假胜地

New Magenta Shores Resort

撰文：Duncan Bainbridge   图片提供：Brett Boardman   翻译：张璐

由 Mirvac 房地产公司开发的马真塔海滨度假胜地位于滨海半岛的海岸上，绵延 2.3km，占地面积为 1.02km^2，是集居住、高尔夫和度假为一体的综合房地产项目，荣获了由澳大利亚景观设计师协会颁发的 2008 年度新南威尔士州设计奖。马真塔海滨度假胜地依偎在太平洋和塔格拉湖之间，恰好在新南威尔士州中央海岸入口以北。中央海岸是澳大利亚新南威尔士州最富活力、发展最迅速的沿海地区之一。驱车从悉尼前往此处仅需 60 分钟，并与沿海小镇 Avoca 和 Terrigal 毗邻。

马真塔海滨度假胜地风景如画——装饰一新的客

房宽敞明亮、设施齐全，还有一室、两室和三室的开放式别墅，从别墅即可将高尔夫球场、泻湖风格的泳池以及滨海公园的景色尽收眼底。

HASSELL 设计集团根据原有的总体规划对马真塔海滨度假胜地的公共空间进行规划设计。设计师综合考虑了街道景观、公共场地以及高尔夫球场的设计样式，使其与周围的环境相适应。设计对象包括所有街道景观、住宅和生活福利设施之间的人行道、一座可供当地居民和游客休闲的小公园以及公园边缘与高尔夫球场相接处的植物种植。马真塔海滨度假胜地的建筑与周围景观浑然天成，这也正是度假村取得成功的关键。

马真塔海滨开发项目在历经重重考验后，终于取得了圆满成功。该项目所在地原本是采沙场，污染侵蚀严重、杂草丛生、自然条件恶劣，并日益受到盐风侵蚀和土地沙化的威胁。于是，人们在这里种植了种类繁多的海滨植物，既有本土的又有来自其他国家的，颜色鲜明、郁郁葱葱。该项目在设计上另一处引人注目的地方就是对水资源进行了可持续监管——监控、滞留和储存优质水源以达到再利用的标准。

马真塔海滨开发项目包括 10 个住宅区，共计约 360 栋房屋，分布在设施完备的 18 洞高尔夫球场和一家名为 Quay West Resort 的五星级酒店的周围。马真塔海滨度假胜地是新南威尔士中央海岸惟一一处私人高尔夫球场，会员及来宾们有机会站在太平洋岸边上使用最新、最完备的设施来一展身手。这座 18 洞的海滨标准高尔夫球场由著名的高尔夫球场设计师 Ross Watson 设计完成，最近刚刚被列为澳大利亚最优秀的高尔夫球场之一。马真塔海滨高尔夫球场球道很宽，而且周围没有传统球场千篇一律的树林、小溪和湖泊，在这里打球可以感受到从塔斯曼海迎面吹来的习习海风。球道旁边是天然的洼地，球场四周有深草区、原始林区以及大约 70 个沙坑。球场的布局和设计不禁使人想起英国与爱尔兰许多著名的高尔夫球场。球场内有两条高飞球弧线，每一条沿球线各有 9 个球洞。前 9 个球洞分布在海边，一边可以欣赏到太平洋的壮丽景色，另一边则可以俯瞰塔格拉湖；后 9 个球洞沿雨林和 Wyrrabalong 国家公园分布，当高尔夫球选手兴致勃勃地打完第 18 个洞时，宏伟的高尔夫球会馆便展现在眼前了。高尔夫球场还为会员准备了完善的练习设施，包括轻击区、准备练习场和障碍练习区等。

马真塔海滨度假胜地由 Mirvac 房地产公司下属的设计公司进行设计，包括一个高尔夫球会馆、3 处游泳池、蒸汽室、一个拥有 5 间美容护理室的水疗中心(其中包括特色维琪浴以及一间情侣专用护理室)，此外还有五星级餐厅、健身房、3 间酒吧、3 个不同用途的会议室、2 间董事会会议室以及 149 栋度假别墅，人们将这些度假别墅称为"沙滩小屋"。

这些别墅的建造为 HASSELL 设计集团的景观设计成果锦上添花，同时也营造出一种舒适、休闲、自然的海滨生活情调。

该项目设计时遇到的最大困难就是如何改造条件艰苦的海岸环境以满足客户高"绿量"的要求。选择植物的时候，尤其要选择抗风、抗侵蚀和耐旱的植物，同时也要兼顾色彩和层次的搭配，使环境符合高尔夫球场、度假胜地和本地居住的要求。

澳大利亚景观设计师协会对该项目的评价是这样的："马真塔海滨度假胜地在园艺景观设计和植物造景方面都有突出表现，对公共空间的设计则充分展示了该度假胜地和住宅区开发的特点。人们祝贺马真塔海滨度假胜地的成功建设，HASSELL 设计集团也在开发海滨度假胜地项目的专业设计方面树立了杰出榜样。"

2008 年已经建成了高尔夫球场、会馆、3 个居住区（其中包括约 96 栋房屋）、1 号度假区（约 80 栋别墅以及游泳池和网球场）、公园和街道景观、海滩步行路以及停车场等建筑设施。目前，正在建设 2 号度假区（包括约 54 栋海滨别墅、开放的娱乐区以及游泳场馆）以及近 29 栋住宅。之后还会按计划兴建另外 6 处居住区，共计 200 栋房屋。

Developed by Mirvac, the award winning Magenta Shores is a unique integrated 102 hectare golf, housing and resort estate located on a 2.3km peninsular beachfront setting. (It was awarded the Australian Institute of Landscape Architecture (AILA) New South Wales Award for design in Landscape Architecture 2008). Nestled between the Pacific Ocean and Tuggerah Lakes, it is just north of The Entrance on the NSW Central Coast, one of NSW's most vibrant and fastest growing coastal areas, 60 minutes drive from Sydney, close to the nearby coastal towns of Avoca and Terrigal.

Set amongst landscaped grounds, the resort consists of spaciously appointed studio guestrooms and 1, 2 and 3 bedroom open-plan villas with views of the golf course, lagoon style resort pool or coastal gardens.

Following an existing Master Plan HASSELL undertook the public domain design of the project, designing and documenting the entire site to integrate the streetscapes, parklands and the golf course into a strongly defined landscape that 'reaches out' to the surrounding area. The design included streetscapes, pedestrian laneways between houses and amenities, a small park for residents and guests and the perimeter planting including interfaces with the golf course. The articulation of the buildings and surrounding landscape is well integrated on the site and contributes strongly to the overall success of the resort.

The Magenta Shores development project is a real triumph over adversity. The site was formerly a contaminated and degraded, weed infested sand mining site, along with the added constraints of salt laden wind and sandy soil. The site has been transformed with the use of a diverse coastal planting palette; a diverse mix of local and exotic plant material provides strong colour and texture, enhancing the lush qualities of the planting scheme. The design is also notable for its achievement of sustainable management of water, high water quality control, retention, and storage as well as reuse benchmarks.

The development comprises 10 residential precincts totaling approximately 360 dwellings that weave in and around a stunning 18-hole golf course and five-star Quay West Resort. Magenta Shores is the only private golf course located on the NSW Central Coast. Members and their guests have the opportunity to play one of the newest and most highly ranked courses in Australia, right along the edge of the Pacific Ocean. This eighteen-hole links style championship course, designed by golf course architect Ross Watson, was recently ranked as one of the best golf courses in Australia. With no traditional lines of trees, streams or lakes, the fairways are wide and open to the winds blowing in off the Tasman Sea, with natural swales and bounded by rough, bushland and around 70 bunkers. The layout and design of this course is reminiscent of many great courses of Britain and Ireland. It consists of two loops, each of nine holes. The front nine starts off overlooking the Pacific Ocean and on the other side Tuggerah Lake. The first nine holes are beach holes with magnificent ocean views. The back nine continue alongside rainforest and the Wyrrabalong National park, finishing with a challenging and interesting eighteenth that brings the golfer to a magnificent clubhouse. This course has a good practice facility, consisting of putting greens, a warm-up driving range and chipping green with bunkers.

The resort itself, designed by Mirvac Design includes a club house, three swimming pools, steam room, day spa with 5 treatment rooms including a specialised Vichy shower and an exclusive couples room, five star restaurant, gym, 3 bars, 3 different conference facilities, two boardrooms and 149 "beach bungalow" resort villas.

These villas have been designed to complement the HASSELL designed landscape and reflect the relaxed nature of a coastal lifestyle.

One of the significant challenges on this project was to deliver a landscape that met the client's high "green volume" planting requirements within a harsh coastal environment. The plant selection consists of species that are particularly wind, salt and drought tolerant, but also provide an array of colours and textures and create a series of stunning settings for golf, resort and residential life.

The Jury Citation from the AILA awards said of the project; "Demonstrating excellence in horticulture and planting design, the quality of the public domain is

a defining characteristic of this resort and residential development. HASSELL are to be congratulated for their excellent competence in this project and have provided a strong example to the profession of design quality in resort developments."

Work on the golf course, club house, 3 residential precincts (approx 96 homes) resort area 1 (approx 80 villas and pool area, tennis courts), parks and streetscapes, beach walk and car parks was completed in 2008. There is at present construction on site of resort area 2 (approx 54 resort villas, open recreational areas and pool complex) plus approximately 29 residential properties still under construction. When these are complete, there is still scheduled design and construction of another 6 residential precincts to include an additional 200 houses.

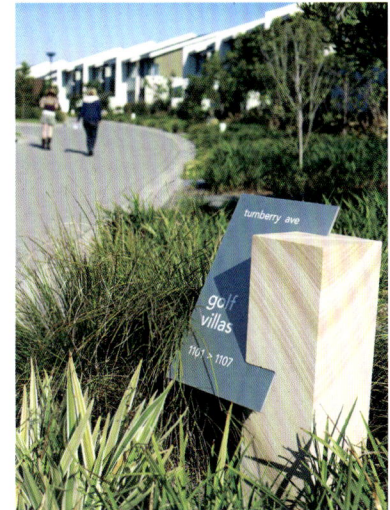

石梅湾艾美度假酒店

Le Meridien Shimei Bay Resort

撰文：Armando Casibang　　图片提供：易道公司　喜达屋酒店　海南华润石梅湾旅游开发有限公司　　翻译：潘亚妮

1　中央景观轴线
2　传统风格的铜鹿雕塑取材于海南当地的神话传说

至今我仍对八年前参观石梅湾艾美度假酒店基地的情形记忆犹新：海滩铺展了上百米宽，入海口在海岸线上形成了一个小缺口，阳光倾斜而下，没有一棵可以遮阴的树木；微风轻拂海浪，拍打着水岸……

早在 2003 年，易道公司受华润集团的委托，为石梅湾 1200 万平方米的新度假社区提供景观总体规划，旨在创造一个全新的生态旅游度假胜地，同时保护和传承原有的自然环境的精华。

该项目位于石梅湾的小海湾中，东侧和北侧由林木茂密的群山环绕，南侧是原始的白沙滩，风景极其秀丽。平地面积约 13 万平方米，其中近 80% 是沙地，零散分布着岩石。当地建筑规范要求所有构筑物的地平面必须比海平面（即百年一遇洪水标高）高出 5m，所以平坦的基地、微小的高差就成为了设计的一大难题。为了满足这一要求，必须建造防洪海堤，基地一大半的区域需进行回填。

总体规划巧妙地利用基地的自然特质，合理安排各种分区的功能。景观设计方案的首要概念是在自然与人类之间达成平衡。前来度假的客人首先追求的是亲近自然的感受，同时要享受五星级度假区的奢华。该设计意在通过提供不同尺度的、丰富的景观环境以及从总体空间规划到精确的细节设计提升客人的自然体验。

迎宾区是度假区的核心，酒店客房集中在东侧，其他设施则位于西侧。10 座带有私家游泳池的别墅坐落于海堤之上，在这里沙滩海景一览无余。别墅后面是双层复式房屋，在二楼可眺望海景。总统别墅位于最私密的区域，并拥有宽敞的海滨花园。水疗设施位于基地东端，设有独立的对外开放的入口。

客人到达迎宾区即能体验到酒店的不同凡响。茂盛的青皮树山林作为背景映衬着酒店，客人从入口坡道进入酒店、穿过迎宾大道到达迎宾广场，当地传统的特色装饰为总体景观拉开了序幕。迎宾广场呈简洁的长方形，采用深色鹅卵石铺装，并种植了茂密、艳丽的植物，摆放着大型的装饰性树池，中央的线性水景由特制的灯笼饰边镶嵌，而广场中央的海枣树把视线引向天空。

酒店大堂是景观设计师、室内设计师和建筑师多方通力合作的成果。各方秉持着相同的意愿——创造

总平面图

能提供真正度假体验的酒店大堂。大堂并不局限在室内，而是与户外空间连成一体，光滑的大理石铺装和水池都延伸到户外。方形水池两边种植着成排的海枣树，加之水池中岛形雕塑相互配合，与远处的家井岛遥相呼应。

最具视觉冲击力的景观是中央景观带。教堂式的屋顶犹如画框，把令人屏息的海景和不远处的家井岛完美地呈现在客人眼前。长条形景观带和迎宾静水池延伸至海面，将客人的视线吸引到自然环境中去。静水池既是户外空间的焦点，也是酒店餐饮、娱乐的核心区域。

从宽敞的花岗岩台阶拾级而下，长方形水池静静地躺在眼前，水池边缘隐没在远处，彷佛与大海相连接。到了晚上，水池闪耀着绿松石的光芒，映照着池边海枣树的叶子。传统风格的铜鹿雕塑取材于海南当地的

神话传说，水池两边摆放着灯笼。从这里回望，中央景观带和静水池成为前景，台阶将人们的视线引入大堂，山脉勾勒出背景，山脊线被优雅、平衡的屋顶打破。

客房和别墅区周围环绕着茂密的天然植被，与迎宾区风格形成鲜明对比。这里成群种植着棕榈和其他树木，呈现出热带花园轻松、安静的氛围，与迎宾区的规则式布置迥然不同。阶梯式种植池和自然式岩石、沟壑营造出不同的效果，解决了基地的高差问题。设计师通过综合运用灌木、树木的分层布置和单个品种植物的群植，成功地为酒店景观创造了丰富的空间，私密的别墅花园或口袋公园都遍植蜘蛛兰。该项目的景观设计更关注细节，例如景墙和种植池的设计，将海南常见的火山岩与切割整齐的砂岩、花岗岩板放置在一起，呈现出有趣的对比效果。

形状不规则的游泳池边环绕着酒店楼房和别墅，

隐藏在葱郁的热带植物之间。水池中设置了一些口袋空间，如在水中设置的坐椅，供私密交谈使用。砂石铺地的泳池甲板上也设置了许多口袋空间，布置着户外家具作为休闲区。有趣的雕塑点缀着水池，增添了度假气息；上岸扶手区独特的茅草屋顶绝对吸引眼球。一楼的套房拥有独立的按摩水池，与游泳池间由填满岩石和卵石的荷塘隔开，水池与荷塘融为一体，可在其中与自然进行亲密的对话。一楼客房的私密性是最重要的，所以设计要保证将私密花园和公共区域之间充分隔断。夜晚，游泳池相当引人注目，绿松石色的水面倒映着摇曳的灯光，仿佛点燃的烛火在水中晃动。

该项目的定位非常独特，要把当地艺术、工艺和传说融入到设计元素之中。设计师根据场地的特质对度假区进行设计，展现着当地的自然景观和文化特色，赋予所有设计元素以场地感。

I still vividly remember visiting the site for the first time some eight years ago. The beach stretched a hundred meters inland, the estuary that broke the cove was a pleasant aberration. The sun was beating down but there was not a single tree in sight for shade. The pleasant breeze seemed to fan the waves that swept the shore.

EDAW's work in Shimei Bay in fact began a few years before this first visit when we were asked by China Resources in 2003 to prepare the overall landscape master plan for the new 1200 hectare resort community. The vision was to create a new eco-tourism resort destination that celebrates and preserves the essence of the existing natural environment.

Nestled in a cove in Shimei Bay, surrounded by densely forested mountains in the north and east, and the wide pristine white sand beach in the south, the Le Meridien site is simply breathtaking. The flat site is approximately 13 hectares of almost 80% sand with a few rock outcrops. The flatness and low elevation of the site became a significant challenge for the design team because the local building ordinance requires that all structures be built above the 5 meter sea level elevation which is just above the 100 year tidal surge level. In order to satisfy this requirement, a seawall had to be engineered and a good 50% of the entire site back filled.

The master plan was quite ingenious in that it took

1 白沙滩上质朴的遮阳伞
2 延伸至海面的迎宾静水池
3 中央景观带夜景
4 夜光下泛着绿松石光泽的不规则水池

advantage of the site's natural attributes in the way the different components were sited. One of EDAW's primary landscape design concept is to express the dichotomy between nature and man and achieving the delicate balance between them. Guests come to Le Meridien primarily to be close to nature, while still be able to enjoy the luxury trappings of an exclusive 5-star resort. As such, EDAW's design intentions were to enhance guest's experiences with nature in a variety of landscape settings and at varying scales, from the overall spatial planning down to the smallest details.

The lobby pavilion naturally became the core of the resort, with the hotel suites concentrated in the east wing and the hotel amenities in the west. There are 10 villas with their own private wading pools, all of which are perched above the landscaped seawall giving them unobstructed views of the beach. Behind the villas are two storey duplexes allowing for ocean views from the upper floor. The presidential villa is tucked away in the most exclusive part of the site where it has an expansive seafront garden. The spa is situated in the eastern most side of the site with its own public entrance.

The dramatization is already apparent right from the arrival experience to the hotel. Against the backdrop of the lush "Qingpi" forest mountain on one side, the guests are greeted at the arrival ramp by three rows of elegant date palms leading into the main arrival plaza. There are two portals at each side of the entry with ornamentation inspired by local folklore. This is a prelude to what is to be the unifying theme of the landscape. The arrival plaza is a simple rectangular courtyard with lush colorful planting, oversized ornamental planters, a central linear water feature lined with custom designed lanterns. The bosque of date palms at the center of the courtyard draws the eyes skyward.

The atrium lobby was a product of well coordinated design collaboration between EDAW, the interior designer and the architect. At the onset, everyone shared the same vision of creating lobby that reflects a true resort arrival experience and in this case a lobby that is both indoor and outdoor, the creamy marble paving of the lobby and the reflecting pool extending into the outdoors. The simple rectangular pool with a row of date palms on either side is broken only by island-like sculptures in the water that echo Jiajing Island in the distance.

One of the most striking features of the landscape and definitely the most photogenic is the central spine where guests are rewarded upon arrival with a breath-taking view of the sea and nearby Jiajing Island that is framed by the porous cathedral ceilinged atrium lobby. The long main spine and the ceremonial reflecting pool in the middle extends out to the sea, fixates the guest's visual focus to the hotel's natural setting. This same reflecting pool becomes the focal point of the next outdoor room which functions as the dining and entertainment core of the hotel.

As the guests descend further down through a series of wide granite stairs, the formal pool emerges from view, seemingly connected to the sea, deceiving the eyes with its infinity edge. At night the pool glimmers a bright turquoise glow bouncing off shimmering reflections on the fronds of the date palms that flank it. A pair of stylized bronze deer sculptures that are inspired by local Hainan mythology, and lanterns flank the jacuzzi overlooking the beach. Looking back towards the mountain from this point, one sees the iconic view of the central spine with the reflecting pool in the foreground, the grand stairs leading up to the lobby pavilion framed by the mountains in the background, and the ridge line broken by the elegantly proportioned roof.

From the arrival spine, the guests will retreat into their hotel rooms and villas. Set amidst lush, naturalistic planting, the guestroom blocks and villa areas contrast strongly with the arrival spine. Here, the palms and trees are clustered groupings contrast with the formal arrangements at the arrival spine, lending a more relaxed and calming ambience to the tropical garden setting. The dichotomy between nature and man is also manifested in other forms, some obvious, and others with more subtlety. In negotiating the level differences within the site, the use of terraced planters versus the more naturalistic rockworks and berms create different effects. Complemented by the use of either a more deliberate layering of shrubs and trees or by the mass-planting of a singular species of trees, EDAW is able to successfully create a rich variety of spaces within the hotel landscape, be it a manicured private villa garden or a natural pocket garden paradise with wind-swept Spider Lilies. On a more subtle scale, such as in the design of feature walls and landscape planters, the textured volcanic rocks ubiquitously found in Hainan, are deliberately juxtaposed with the more refined sawn-cut sandstone and granite slabs to create an interesting, contrasting effect.

Carved into a space surrounded by the hotel towers and two-storey villas is a free-form swimming pool. Lush tropical planting screens the adjacent guest rooms and villas from view. Within the pool are several pockets with in-water seating for intimate gatherings. Likewise,

there are plenty of pocket spaces around the sandstone paved pool deck for lounging and furniture setting. Interesting sculptures dot the pool area adding to the resort atmosphere. The swim up bar cannot be missed with its tall thatched roof. Guest rooms on the upper floors overlook the pool while the ground level units each have their own jacuzzi pool. Separated from the main swimming pool by an interconnected system of lily ponds with rocks and boulders, the jacuzzi pools were designed to become one with the pond where the experience is truly communing with nature. Privacy for the ground level units is a priority so EDAW made sure there is always sufficient buffer between the private gardens and the public areas while framing vistas at every opportunity. At night the freeform pool is at its most dramatic with the gentle glow of the turquoise green water and the flickering pool lights that appear to be candles burning underwater.

At the Le Meridien, we envisioned to make this experience truly unique by incorporating design elements inspired by local arts and crafts and folklore. EDAW believes that the resort experience should be site specific, showcasing the location's natural and cultural resources, with all the design elements helping to celebrate its sense of place.

1　入口广场中央的海枣树把视线引向天空
2　附带独立泳池的别墅一楼客房
3　连接户外景观与室内的酒店大堂

阿塔卡马沙漠中的生命线 —— 普利塔玛地热温泉

A Life Line in the Atacama Desert — Puritama Geothermal Springs

撰文：Jimena Martignoni　　图片提供：Guy Wenborne　Jimena Martignoni　　翻译：丁岩

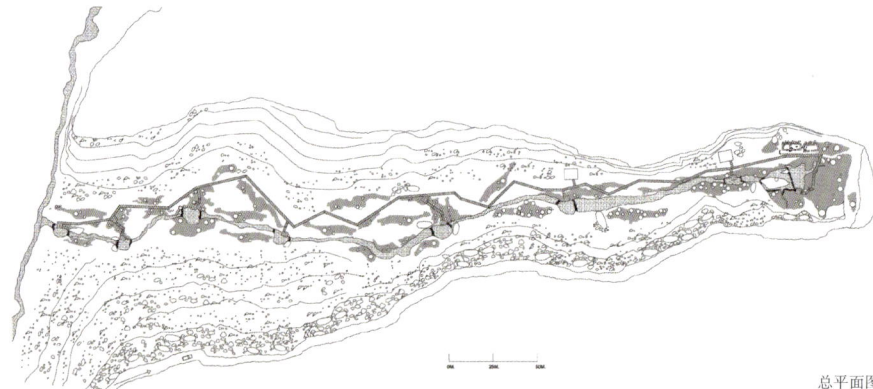

总平面图

普利塔玛温泉位于智利，距圣佩德罗 – 德阿塔卡马镇 33 796.29m，该镇是阿塔卡马沙漠海拔 2438.4m 处的一片绿洲，源于玻利维亚高原冬季的降雨。在一望无际的呈红色、赭石色或褐色的贫瘠山脉间（山脉颜色依光线的变化而变化），圣佩德罗 – 德阿塔卡马镇仿佛是矗立在沙漠中的一座孤立的绿岛。

这座小镇最初由阿塔卡马人建立，是一个受玻利维亚的艾马拉印第安人影响较多的农业部落。在 1425 年被印加人占领，又于 1535 年沦为西班牙的殖民地。如今，阿塔卡马人的后代仍保留着一些传统文化，分布在沙漠上具有医疗作用的温泉便是其中之一。

圣佩德罗 – 德阿塔卡马镇目前是智利接纳来自全世界的观光游客最多的城镇之一，旅游业是该镇主要的经济支柱，该项目也成为了该镇吸引游客的主要景点。大部分智利人民还不具备完全成熟的环境意识，但小型考古城镇的文化和自然遗产已经开始受到关注。考古学家、圣地亚哥前哥伦比亚博物馆馆长 Carlos Aldunate del Solar 是这个景点的特别顾问，他与建筑师 German del Sol（主创设计师）共同合作设计该项目。

在圣佩德罗 – 德阿塔卡马镇，耕地采用轮流灌溉的方法；公共蓄水池——当地河流通道系统的一部分，严格地按照轮流时间表为耕地提供灌溉用水。对于当地人和游客们来说，北方沙漠中的地热温泉是不可多得的天赐之物，成为地球上最干燥地区的一道生命线。

智利地质差异很大，北部有世界上最干旱的沙漠——阿塔卡马沙漠，南部为茂密的巴塔哥尼亚森林和蔚蓝湖泊，在这处大约 4 345 千米的范围内具有多样

的生存环境。然而，这里有个突出的地质特点——安第斯山脉，连接了从玻利维亚高原到火地岛的许多智利景观。

作为山脉的一部分，地热温泉分布在不同的区域，主要集中在三个地区——北部高原地区、中央区及智利南部。

泉水是从特定区域流出的地下水，而温泉或地热温泉则是水温远高于当地年平均气温的泉水。当雨水降落到周围高地上时，它渗透到含有矿物质的岩石中，并在地球内部进行加热。当接触到大的断层或裂缝时，被加热的水就顺着裂缝上涌到地表，形成温泉。为了使温泉水在流出前保持一定温度，水必须通过直接的通道快速到达地表，而这个通道是在特殊的石灰岩层中天然形成的。

普利塔玛地热河从 304.8m 深的沙漠岩石裂缝中喷出，流出时的温度为 30℃~33℃的。普利塔玛（puritama）这个名字最初就是由水（puri）以及热（tama）组成。主创设计师初次参观此处时，这条河被安第斯蒲苇（一种当地的观赏植物）所覆盖。沿线的一些温泉是在古时就被发现并一直保留下来的。景点入口处保留着印加人攻占智利北部时建造的两座建筑物，成为象征远古的印记。

在阿塔卡马 Explora 酒店买下这块场地后，主创设计师出于对其公共用途的考虑，设计了这个方案。普利塔玛温泉的建设只用了不到 3 个月的时间，但是滴灌系统的安装及印加棚屋的修复延长了项目的完成时间，最终于 2000 年对公众开放。

设计和建设都是完全在原场地进行的，设计团队在现场用绳子测量木栈道的精确位置，然后再将数据体现到纸面上。

设计师拓宽并加深了 8 个池塘，其中部分池塘周围建有石墙。由于新增了滴灌系统，安第斯蒲苇十分茂盛，将石质建筑完全遮蔽。根据光照位置的不同，银色的蒲苇在一天中可以形成不同的光影效果。

主入口处有两个简易的白色盒子状的建筑物，里面设有洗浴室、更衣室和桑拿房，可供游客使用，建筑物的用水也是由提供灌溉用水的蓄水池提供的。

经过修复后的印加棚屋变成了管理办公室，新增的茅草屋顶保留了原建筑的风格。整个场地由距离河面约 0.3 米高的木栈道连接，这样既能使木板下的植物茂盛生长，也为游客穿越各个温泉池提供了便利。木栈道到达温泉池时变得宽阔起来，游客可以享受日光浴或者在此休息。

走近场地与穿过场地的感受可谓完全不同。从山坡的小路上可俯瞰全景：一条细长的木栈道在广袤的银色蒲苇丛中若隐若现，木栈道被漆成暗红色，与闪烁着银光的蒲苇及周围的山脉形成鲜明的颜色对比。

只有当人们走近场地的停车场时，这些分散的温泉池才显现出来。停车场被设置在高原上，与场地保持同一高度。人们将车子停在此处，并从这里踏上木栈道的入口，但道路却被高大的安第斯蒲苇遮蔽住而变得不清晰。

实际上，只有走在温泉池旁的木栈道上才能发现温泉以及由木栈道形成的平台。有些温泉完全被蒲苇包围住了，显得十分孤立，景色也仅限于蓝天和周围的群山；其他被蒲苇包围的温泉景色则更开阔些，池水十分温暖，绝大部分的温泉都有小瀑布，起到了自然按摩以及淋浴的作用。

The Puritama thermal baths are located 21 miles from San Pedro de Atacama, an oasis 8,000 ft above sea level in the Atacama desert, in Chile, originating from the rainfall of the invierno altiplanico de Bolivia or Bolivian altiplano winter. In the middle of an endless landscape of arid mountains colored in red, ochre and sepia (depending on the morning, noon or evening light), lies in the town of San Pedro, which from the distance emerges as an isolated green node of life.

This city, first settled by the atacame~os—a local farming culture influenced by the aymaras pueblos in Bolivia—was conquered in 1425 by the Incas and then colonized by the Spaniards in 1535. Today, some of the costumes and culture of the atacame~os are saved by its descendants, such as the use of thermal waters with curative effects scattered along the desert.

As San Pedro is nowadays one of the most visited Chilean towns by tourists from all over the world, and tourism is the main economic activity, this costume is also preserved as one of the main attractions of the area. While the Chilean community at large has not a completely mature environmental awareness yet, these kinds of small archaeological towns are starting to take good care of their cultural and natural patrimony. Archaeologist Carlos Aldunate del Solar, director of the Pre-Columbian Museum in Santiago, was the specialist consultant at the site who worked hand-in-hand with architect German del Sol, the chief designer.

In San Pedro de Atacama farming areas take turns to be watered; common cisterns, that are part of a system of channels fed by local rivers, provide water under a very rigid schedule that everyone respects. The thermal baths in the northern desert are an unexpected gift much appreciated and cared for by locals and visitors: a subtle line of water and life hidden in one of the driest places on Earth.

Landscape diversity in Chile varies from the world's most arid and driest desert, the Atacama Desert in the north, to the exuberant Patagonian woods and blue lakes in the south, creating along the country's extreme length of approximately 2,700 miles, very different environments in which to live. However, there is a dominant physical feature which extends along the entire length, from the Bolivian plateau to Tierra del Fuego, which visually and geologically connects the many Chilean landscapes: the Andes.

As part of this mountain system, geothermal water sources or hot springs are dispersed throughout the country's different regions, mainly concentrated in three areas: the northern high plateau district; the second district in the central region; and the third district in southern Chile.

Springs are places where groundwater is discharged at a specific location. Hot or thermal springs are defined as springs where the temperature of the water lies considerably above the annual air temperature of the region. As rain falls on the surrounding peaks, it infiltrates the rocks picking minerals up and heating up from the primal heat of the Earth. When it encounters a large fault or crack the now heated water ascends along the fault-line to the surface as a hot or warm spring. For the water not to cool back before it bubbles out, it has to follow a direct, fast route to the surface, which is assured by the pipelines naturally carved by the water itself, within the typical limestone formations of the rock.

The Puritama geothermal river flows along an almost 1000 ft-deep crack carved in the rocky desert mountains,

bubbling out as 86 to 92 F-temperature springs. The original name of Puritama actually comes from puri that means water, and tama that means hot. When German del Sol visited the site for the first time, the river's borders were covered with Andean pampas grass or cola de zorro (Cortaderia atacamensis), a highly ornamental native plant. Along this linear natural system, some pools of hot water were left from past times when local people discovered them. Also reminiscent of the past, when Incas conquered the north of Chile, two constructions completely built in adobe (kayankas) lie near the site's entry point.

This scenario is what German del Sol worked with and modeled for public use, after the land was bought by the Atacama's Explora Hotel. The construction of Puritama took no more than three months, but the installation of the drip irrigation system and the restoration of the Inca shelters, which was made in phases, delayed the finishing. The site was opened to the public in 2000.

Both the design and construction were one-hundred percent in situ processes, in which German del Sol, together with other people from his office, worked at the site marking the boardwalk's exact placement with ropes, then to be translated onto paper.

The eight pools were broadened, deepened and part covered with stone walls. As a consequence of the drip irrigation system that was added by the designer, the Andean pampas grasses now grow exuberantly in thick masses that hide the stone structures. Furthermore, their silver crests become light-attracting elements generating a mystical effect during certain times of the day, and creating diverse light situations depending on the sun's position.

Two minimalist-looking white box-shaped constructions stand close to the main access, each of them offering a bathroom, a dressing-room and a sauna, which are much

1 该项目的主要元素：本土植物
2 该项目的主要元素：温泉
3 温泉池旁的平台

used by visitors either before or after they get into the hot springs. The running water is provided by the same small collecting pool that feeds the irrigation system.

The kayankas, or small Inca shelters, were restored and rehabilitated as administration offices, and new thatched roofs were added reminding of original constructions. The site is entirely crossed by a wooden boardwalk elevated approximately one foot over the course of the river, which allows the grass underneath to grow while providing a walking path to access the different pools and resting spots. When the floating boardwalk reaches the pools, it widens up and creates small terraces that face the water where people lay down to sunbathe or just to rest.

The experience of the site varies as one is approaching it and gets much different when one is actually walking through it. Seen from the road that goes down to the mountain slopes, the site is presented as a whole; shaped as a green strip of tall grasses crowned by a hundred white-silver crests, and crossed by a long, narrow path that gets lost in the vast arid landscape. The path, painted in dark red, creates a mayor color contrast with the swinging luminous crests and with the sepias and grey hues of the enclosing mountains.

The scattered pools where people gather become noticeable just when one is getting closer to the site's parking area, which is placed in a plateau at the same level as the site itself. From here, one leaves the car and steps into the boardwalk entry point without knowing the exact direction this could take, for the tall Andean pampas grasses screen the many turns it makes, as well as each one of the pools.

One literally discovers the pools and the balconies shaped by the boardwalk while walking along it. Some are completely surrounded by the Andean grasses, becoming more isolated, intimate spots from where one's view is reduced to the blue of the sky and the surrounding mountains; and others are framed by the grasses in a more open manner creating wider vistas of the site. Inside the pools the water is warm; the small waterfalls, which most of them have, act as natural massage showers and the sun feels strong (solar radiation in Atacama is really high).

遥远的景观，遥远的美丽 —— 拉丁美洲Remota旅馆

Remote Landscapes, Remote Beauty — Remota Hotel in Latin American Land

撰文 / 图片提供：Jimena Martignoni   翻译：李沐菲

Remota 旅馆坐落在南美洲的安第斯山脉中，位于 Señoret 运河沿岸，是这一地区独一无二的标志性建筑。建筑物与其周围的景观设计巧妙地融为一体，甚至很难界定二者的界限。

Remota 旅馆堪称是一座宏伟的建筑，以其独特的造型傲立于这片土地之上，其流畅的几何线条体现了人类的奇思妙想与聪慧。远远望去，人们就能感受到这些线条的自然与流畅，营造出多元化的视觉效果。

而使这项工程无论从审美角度还是理念方面均与众不同的原因主要有两个：第一，设计师根据该项目特殊的地理位置中的景观特征来确定建筑的风格；第二，当地的自然与文化景观已成为了建筑的一部分，使人们回忆起这片遥远地域的起源和历史。

项目的所在地早期以发展畜牧业为主。即使是现在，当人们驱车穿过通往"世界尽头"的公路时，也会遇到驱赶着羊群的牧羊人，这时人们便会为其让路，

使羊群可以继续悠闲地漫步。

变幻多彩的天空颜色、静谧的环境、拂面的微风、宜人的田园景观……诉说着 Remota 旅馆这份遥远的美丽。

2004 年，委托建筑师 German del Sol 来设计这座"世界尽头"的旅馆。German del Sol 是智利的一位杰出设计师，他对建筑物周围的景观设计有着独到的见解和感悟，他的多数作品都将景观作为开启灵感的重要因素，而 Remota 旅馆也必然成为这一地区景观特征的产物。

建筑物沿着海岸线自然地伸展，而这一区域作为

草原的一部分也在向远处延伸着。夏季的草原郁郁葱葱，到了冬季则变成了金黄色，与清冷的天空和水面形成了强烈的对比，正是这种鲜明、生动的视觉反差激发了设计师的灵感，并以此进行建筑设计。

设计师不是简单地将建筑置于景观之中，而是将二者完美地融合在一起。这不仅需要明确设计理念（即将周围景观融入到设计中去，而不是再造和复制），更重要的是需要人们以一种诚恳的态度去理解、去保留景观的原貌，因为景观本身是早于所有人为的规划和

建筑而存在的。

为将人们的视线引向最美的景色，设计师将建筑沿海岸线建造是最常规的做法，但设计师认为这一地区还有很多不同的景色吸引着人们。因此，设计师在这片面积为 44 517 m² 的区域里沿着西北和东南两个方向分别设计了两座长条形建筑，而沿着东北方向设计了第三座建筑。为了设计一处中心庭院，设计师沿西南方向建造了一个露天走廊，这样就可以使置身于旅馆之中的人们至少可以欣赏到三个方向的奇美景色，

甚至是周围景观的全貌。积雪覆盖的山峰、浩瀚的海洋和辽阔的草原都是典型的当地景观，设计师利用这三座建筑勾勒出一块不规则形状的广场，面积约为48 564平方米，犹如一片金色的草原，使人们近距离地欣赏到大自然的美景，也使人们更加贴近大自然。

在这一区域增加的惟一元素就是为了体现巴塔哥尼亚的文化风格而放置的大石块，由于其不规则的摆放，因此被称为"乱石"。设计师将这些不规则的石块放置在草地上，使其呈现出一幅迷人的画面——这些质地坚硬的石块沉沉地陷入草地中，纹丝不动，而草叶轻柔地在风中摆动，形成强烈的对比。

设计师为体现出该项目的本土特色，不仅运用了对比的方式，还应用了呼应的方法，力求保持该地区现有的景观元素。建筑立面较深的颜色以及走廊与草地和多云的天空构成了强烈的视觉对比，又与远处连绵的群山相呼应。深夜来临，整栋建筑都隐藏在夜色之中，只有闪烁的梦幻般的灯光体现着它的存在。

设计师直接采用木材作为旅馆建筑的外墙，这也是基于当地牧场式的人文特征。在这一偏远地区，畜牧业曾是最兴旺的产业，也是该地区最有代表性的一种生活方式。因此，第一代土地所有者所居住的牧场

庄园也就成为了辽阔的巴塔哥尼亚地域中最具有象征意义的建筑形态。Remota 旅馆的设计向人们清晰地再现了早期坚固而简洁的木质建筑，更深刻地描绘了这片土地独特的景色。但是最能巧妙地反映出巴塔哥尼亚畜牧业人文特征的设计细节却是连接三座建筑的步道。两条木质步道从广场的东南部一直延伸到西北部，蜿蜒曲折，人们似乎可以联想到羊群被赶进庄园时的情景。更有趣的是，设计师在设计通往旅馆内部各个房间的走廊时也采用了这种风格。

露天的西南走廊将旅馆的中央庭院封闭起来，从海边望过去，走廊如同一道深色的轮廓。当秋季来临时，走廊与金黄色的草地形成鲜明的对比，犹如一道传统的木栅栏。

设计师坚持保留景观的原始形态，并强调屋顶平台只是地面的抬升，因此旅馆的屋顶都经过景观处理。这些屋顶平面不仅促进了建筑与自然的融合，还能起到隔热及防火的作用。从高处俯瞰，所有的屋顶似乎都是地面的一部分，使整个建筑与景观完美地融为一体。

巴塔哥尼亚的自然景观是由粗旷的线条构成的，因而项目也要保留这种美感，力求集合景观原有的形与色。设计师以认真的态度对当地景观的主要风格以及众多其他元素进行了分析和设计，最终使 Remota 旅馆稳稳地坐落于这片草原之上，并力求使一切都达到完美的平衡。

At the furthermost piece of South-American land, facing the seacoast of Canal de Señoret and the very last peaks of the Andean Mountains, rests Remota Hotel: one of those works of architecture whose only presence becomes a landmark and, consequently, the surrounding landscape becomes part of it. When the boundaries between an architectural project and one of landscape architecture become this vague, probably the first question that is fairly raised is really how well-defined they are; or, ultimately, how tricky it is sometimes to determine the connotation of each of them.

Remota Hotel is indeed a piece of architecture, for it appears onto the landscape as a distinct creation whose purely geometrical lines make it obviously a man's work. Yet, a first superficial look would rapidly expose how those lines were thought of never to be perceived as straight, lifeless shapes; they are twisted, sometimes almost curved, and create multiple options for the eye. In a subsequent more thoughtful look, two things which become the main reasons why this project is not just aesthetically but conceptually different come to surface. In the first place, the manner in which the landscape is not just integrated into the project but used to define the "exterior architecture"; here, the outdoors space is the existing landscape itself. Secondly, the many references to the local natural and cultural landscape which, realized as part of the architecture, stand as reminders of the origins and history of this remote region.

This project is located in Puerto Natales, a fishing town in Chilean Patagonia whose origin is deeply connected with sheep rearing. Today, it's still common to be driving through the typical "end of the world" Patagonian roads, isolated and eternal, and being forced to slow down or just stop to let a flock of hundreds of sheep and its shepherd continue on their unhurried way.

The place is remote, as is it's beauty; at the very least, a different distant kind of beauty, which can be appreciated in the changing skies, the silence of the site, only disturbed by the wind, and the bucolic landscape of the "matorrales".

The commission to design a hotel at the world's end was given to architect German del Sol in 2004 by a Chilean private developer group. German del Sol is a remarkable designer in Chile whose work is characterized by a profound respect and understanding of the landscape in which his projects are inserted; in most of them, landscape actually appears as the inspirational leading element that makes the architecture shaped as a mere result of the site's character.

The site was untouched. Naturally developed as a continuous plane that descends towards the level of the sea coast, the land where this project is placed extends as part of the never-ending horizontal fields that characterize the Patagonian steppes. While green during the summer, these grasslands turn into golden surfaces that contrast with the cold tones of the sky and the water in the winter; this dramatic visual change is probably the only really noticeable alteration that this landscape goes through, and the one that motivated German del Sol to use it as a key design element of the project.

Incorporating the existing landscape into the architectural project, instead of just placing the building as a podium from where to enjoy amazing views, was the first objective the designer set. This inclusion, however, is not based on a merely conceptual gesture that brings "the idea" of the enclosing landscape into the design; neither as a recreation nor as a replica. Conversely, in a probably more humble but also more honest manner, the inclusion of the landscape is based on the simple decision of leaving it there, as it appeared before any thought of architecture was even dared.

"Placing the building parallel to the seashore would have been the most obvious decision to direct the vistas towards what one thinks is the best view", says German del Sol, "But the truth is that not only one but many different views at this place are definitely worth being drawn to attention." Therefore, he positioned two linear buildings along the northwest and southeast sides of the 11 acre-site, facing each other; a third one along the northeast side; and, in order to define a central courtyard or "natural plaza", he partially closed the space on the southwest side with an unroofed walkway. In this way, at least three amazing views are possible to be enjoyed from the hotel, thus providing a final more panoramic option to watch the surrounding landscape. However, these are still the typical "distant" views, which bring the snow-capped mountains, the sea and the grassy plains only visually closer to the hotel guests; the decision to trim part of those plains to create a natural plaza, on the other hand, represents the possibility of bringing them physically closer. Shaped by the three framing buildings into an irregular rectangle, this 1.2 acre-central space becomes a single uninterrupted surface of golden grasses, and appears as a more private, defined piece of the original landscape.

The only elements that were added to this area are a number of boulders which, placed among the grasses, offer a cultural reference to the Patagonian setting. Generally known as "erratic rocks" due to their errant location, the geological origin of these stones is related to the displacement of ice glaciers which, on their way, carried them and finally left them thousands of kilometers away from their original position. The stones that are placed in the hotel's central plaza were brought from the nearby fields by the designer and the construction site's workers. The sight of these rough stones settled into the golden-brownish grasses generates an especially attractive picture: the deep contrast created by the steadiness of the boulders, heavily positioned on the ground, and the constant light swaying of the grasses in the wind.

Other contrasting and, in opposition, matching effects were also implemented by German del Sol to define the vernacular character of this project; this definition is therefore based on a constant playing with the elements of the given landscape. The dark color of the lenga wood that structures the buildings' facades, as well as the chimneys and the connecting walkways, creates a markedly visual contrast with the grass and with the usually cloudy sky, but at the same time matches the darkish shades of the distant mountains. At night, the whole construction seems to be hidden in the landscape, and the only sign of its presence is the quite surreal image of long strips of lights that seem to come vertically out of the ground.

The decision to use wood to define the general appearance of the hotel was also based on a very specific reference to the local cultural landscape of the estancias. As part of the livestock activities historically developed in this isolated region and the highly representative lifestyle associated with it, the estancias turn into the most emblematic construction of the extensive Patagonian territories on which the first landowners had to settle. The layout for the Remota Hotel clearly recalls these robust but simple wooden farm buildings that, to the present day, portray this landscape. However, what probably best, or at least more resourcefully, evokes

the vernacular repertoire of the Patagonian sheep culture is the design of the hotel's walkways which connect the three buildings. The direct allusion to the long winding pathways that sheep have to walk to be sheared in specialized estancias is especially noticeable at the main two wooden walkways, which extend along the entire length of the plaza from southeast to northwest. The reason why sheep paths are not completely straight is to guide the animals' progress and make them reach the shearing spot one by one; in a subtle almost funny way, the designer also chose to emulate this singular layout for the hotel's interior corridors that lead to the rooms.

The southwest walkway, which is unroofed and closes the hotel's courtyard, is visually perceived from the coast as a dark contour that crosses the entire site; only perceptible due to the color contrast produced by the yellowish surface of the grasses, this piece can easily be mistaken for any typical local wooden fence.

Remaining true to the concept and emphasizing the idea of raising the roof planes out of the ground, every one of the buildings that make up the hotel features green-roofs. Apart from providing thermal insulation and acting as fireproof surfaces, these grassed-planes turn into another element that helps the construction to match the natural

enclosure. From the highest points of the site, the roofs seem to extend as part of the ground, and the view of the architectural complex becomes one with the landscape.

"This project wants to sum up the colors and shapes of the original landscape", says German del Sol, and then adds: "Patagonia is made of coarse lines…, and this project wants to continue with this aesthetic." Certainly achieved, these goals were set under a very reverential attitude to the many elements and general image of the local scenery, and although the construction presents itself very solidly on the landscape, when weighing one against the other, the result is finely balanced.

道路

古迹重塑 —— 帝国宫殿大桥

Reshaping the Historic Site — Imperial Palace Bridge

撰文 / 图片提供：AUBÖCK+KÁRÁSZ LANDSCAPE ARCHITECTS and ARCHITECTS

翻译：张咏梅

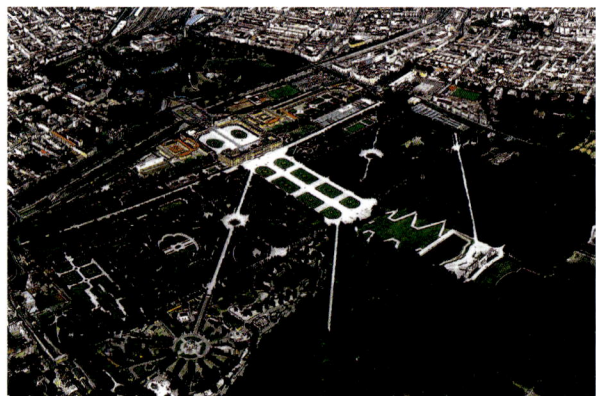

　　该项目是奥地利首都维也纳滨海区大型改造项目中第一个被付诸实施的工程，是由 AUBÖCK+KÁRÁSZ 与 Szedenik&Schindler 两家公司的设计师共同设计完成的项目，并荣获了国际竞赛奖项。该项目重新规划了滨海区，包括地下停车场、中心广场以及其中一个著名景点的全新设计。

　　帝国宫殿大桥有着悠久的历史——1758 年，这里曾建有一座木质小桥，它垂直于城堡的轴心，是当时进入宫殿的主要通道。到了 1900 年，受河流规划的影响，这座小桥进行了扩建，并被改建为石质结构，形成了一个小型现代广场，交通便利。在过去的几十年里，这个小型广场主要被作为停车场使用。

　　设计师试图改变这种不合理用途，将其设计成为具有代表性的公共空间。设计师在广场的中心地带设计了由碎石块铺设成的步行道和三级天然花岗岩的台阶；步行道两侧是草坪，其上栽种的黄杨木呈几何形状排列，体现了传统的轴对称性的特点；位于桥体各个方位的斯芬克斯狮身人面雕塑则体现了 18 世纪的历史特色。

The Imperial bridge project is a first step of realisation after winning an international competition (together with S & S architects) for the foreland of Schönbrunn castle in Vienna/Austria. The entire project foresees a large scale restructuring of the foreland of Schönbrunn by creating an underground garage, an arrival centre and a completely new surface design for one of the most attractive tourist sites of Vienna.

Originally a wooden bridge was built in 1758, erected in the axis of the castle serving as the main entrance to the palace. Around 1900 the river was regulated, the bridge was enlarged and built in stone, creating a modern square dominated by the traffic. In the last decades this square on the bridge was mainly used for car-parking lots.

The new design is based on the elimination of this disuse by regaining public space as a particularly representative site. Now the central strip is dedicated to pedestrians only, covered with gravel surface, positioned 3 steps above a natural stone paving area of granite. The lateral parts are lawned, accentuated by geometrically placed boxwood (buxus) spheres. Thus the traditional axiality is specially underlined. The sculptures of the sphinges and lions mark the dimensions of the original bridge being elements of the historical continuity since the 18th century.

公共艺术大道 —— 刘易斯大道

The Public Art — Lewis Avenue Corridor

撰文 / 图片提供：SWA Group　　翻译：张璐

总平面图 1

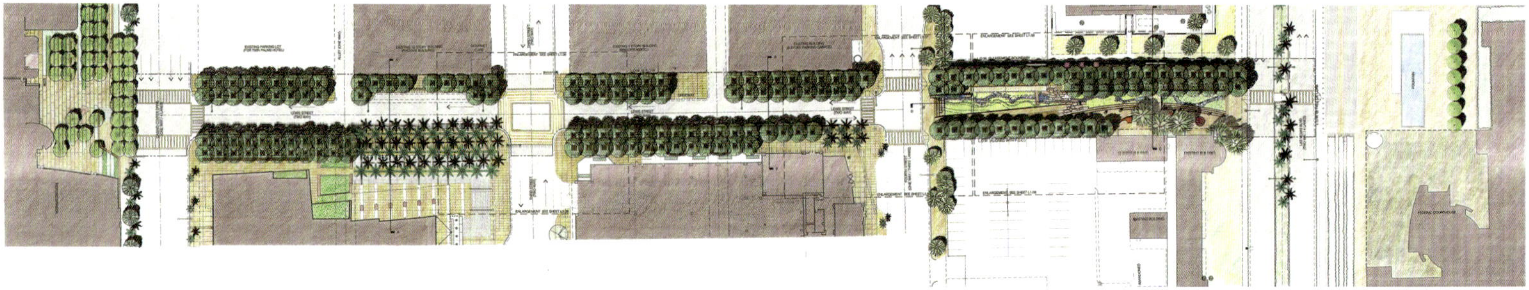

总平面图 2

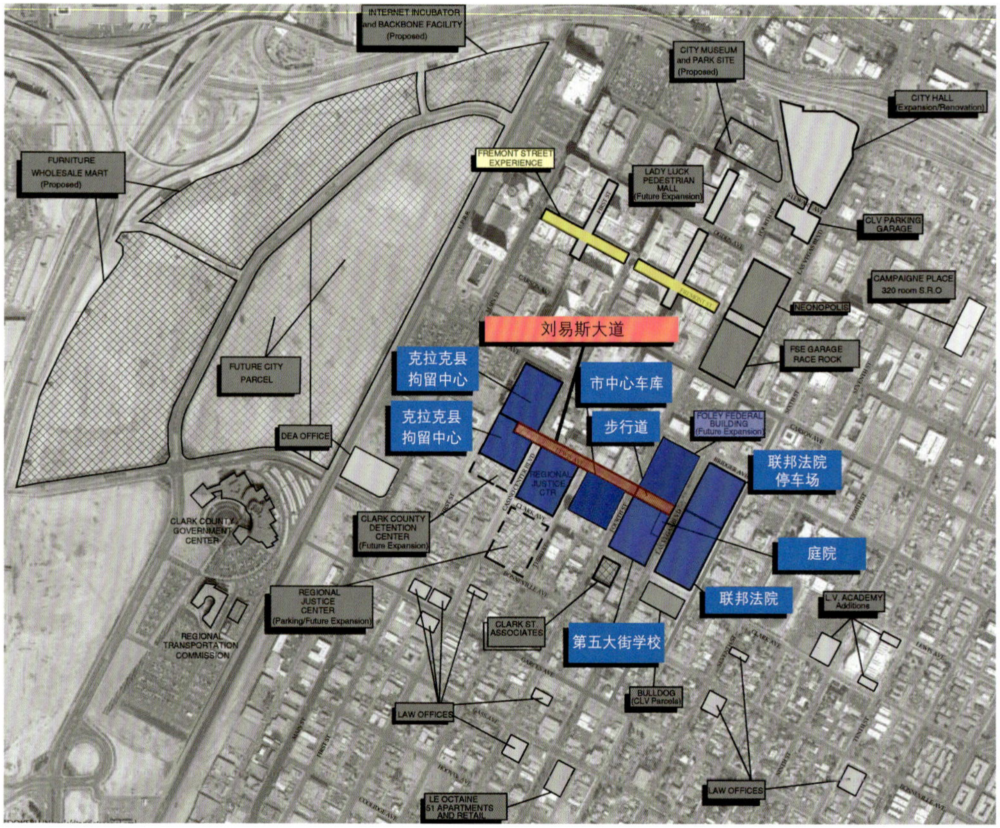

刘易斯大道——一端是克拉克县拘留中心，另一端是联邦法院

　　提起拉斯维加斯，人们首先联想到的便是充斥着刺眼的灯光、诱人的赌博机器和豪华的加长轿车的"大道区"，但它除了是闻名遐迩的赌城，更是拥有1700万人口的城市。从2000年开始，拉斯维加斯市的市长奥斯卡·古德曼便一直酝酿在市中心为市民建造一处"公共空间"，并将其确定为"拉斯维加斯百年规划方案"，旨在使城市面貌焕然一新，成为一个集文化、政治、金融和商业为一体的城市。

　　拉斯维加斯市政府于1995年向联合太平洋铁路公司征用了约246 855平方米的土地，邀请SWA的设计师对其进行设计，希望能够吸引更多游客及当地居民前往市中心。为了实现这一目标，设计师决定在此建造一片开阔场地，不仅可以为行人提供方便，还可以吸引更多市民参与经常在此举办的文化活动，从而促进该地区其他的重建工作。

　　为了使该项目可以顺利实施，社区投入了大量的资金。设计师新建了一条穿越拉斯维加斯市中心的刘易斯大道，其一端是克拉克县拘留中心，另一端则是相隔三个街区的联邦法院的新大楼。大道的两旁还分布着若干历史性建筑物以及市、县和地区行政办公楼，并一直延伸至艺术区和附近的住宅区。

　　在重建之前，项目的场地环境显得十分粗糙：大道两旁的建筑缺少人文气息，而且没有任何景观设施，只是突兀地矗立在行车道两旁；人行道也十分狭窄且完全没有树阴遮挡，因此，很少有行人经过此地。为了解决这一问题，设计师在大道两旁重新划分了人行道和车行道——把四条机动车道中的两条改建为人行道，将人行道扩至 6m 宽。

　　设计师最初计划在联邦法院和地区司法中心的现址上建一条连接公共场地的直线形街道，这样联邦法

院将被建在高出街道台阶 3m 的地方,并可以在大门处建造喷泉景观,由此便产生了街道景观的新亮点——喷泉。

项目的设计始于社区的讨论会,普通市民也被邀请前来参与讨论,拉斯维加斯艺术协会的成员和当地的艺术家们也都纷纷献计献策,于是公共艺术大道的想法便应运而生了。设计师将营造可供行人使用的活动空间放在首位,在拓宽人行道的同时在街道两旁栽种了白蜡树,这样既可以营造出树影婆娑的氛围,也可以缓和周围建筑所带来的生硬感。另外,在地区司法中心和综合使用区周围种植的棕榈树与白蜡树交相辉映。

在街道两旁种植树木是十分重要的环节,设计师计划在树木周围修建排水沟以便雨水能充分渗透至树木根部,并在街道两侧种植了适合在沙漠地区生长的树种,如光棍树和豆科灌木。设计师在低于人行道两级台阶的地方修建了一条布满鹅卵石的蜿蜒小路作为中央水景,犹如一幅置于场地中央的水墨画,诠释着自然与艺术的融合。中央水景逐渐由鹅卵石小路演变成小溪,为层叠的水景墙提供水源。水景墙沿广场中轴线设置,并垂直于第四大道的方向,设计师将很多适合在河边生长的本土植物移植到此,人们的步行道路由暖色调的小桥或天然的石子路组成。

项目所用的材料十分简单,设计师充分考虑了施工和维修成本,因此坐椅和街道两旁的公共设施都使用了暖色调且坚固耐用的材料,并保留了原有的圆形路灯。

项目排水系统的设计也别出心裁。由于拉斯维加斯市的降雨量很小，因此城市几乎无需设置排水系统，但为了避免个别降雨量过大的情况而给行人带来不便，设计师将整个路面倾斜，在较低的一端设置了一个雨水滞留池，以使街道的积水不会对行人造成困扰。

艺术家们纷纷以刘易斯大道为主题展开创作，并赋予其"诗人的公园"这一美誉。由18位艺术家创作的诗歌和散文被刻在步行桥上，成为了刘易斯大道的标志性代表之一。刘易斯大道的建设对重建拉斯维加斯市中心计划产生了深远的影响。人们来到此地都会情不自禁地想要感受一下在大道上步行的乐趣，刘易斯大道为拉斯维加斯市民提供了一个充满活力的城市空间。

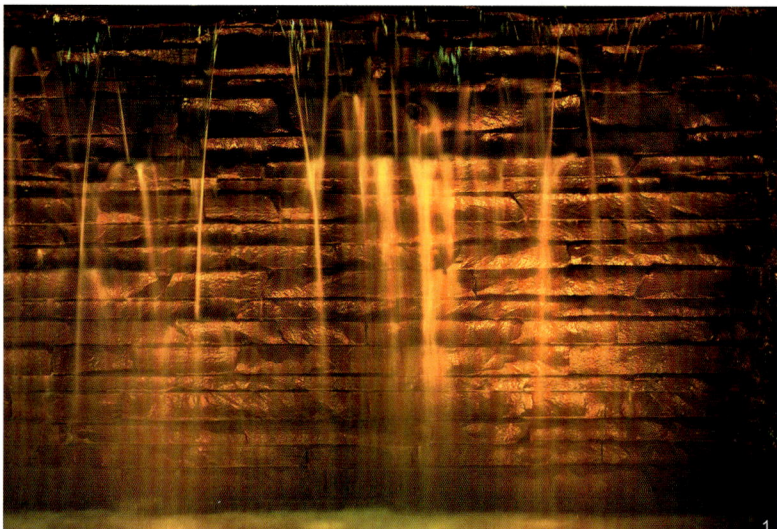

1 水景墙在橙色灯光的映照下营造出静谧的氛围
2 刘易斯大道是人们聚会、举办活动的公共空间
3 街道两旁栽种的白蜡树为行人提供了斑驳的树阴

To many people, the name Las Vegas evokes "The Strip"; bright lights, slot machines, and stretch limousines. But the city consists of much more than the infamous gaming industry; the City is home to more than 1.7 million people. In 2000, Mayor Oscar Goodman began an effort to create a more pedestrian-oriented "people place" in downtown Las Vegas. His development goals for the City were expressed in the "Las Vegas Downtown Centennial Plan" a collaborative planning document intended to change the face of downtown into premier cultural, civic, financial, and business center.

SWA had already been involved in site design for the Federal Courthouse and the Regional Justice Center in the downtown district, so when the City purchased 61 acres of property from the Union pacific Railroad in 1995, SWA was invited to submit designs that would realize the goal of draw more people, both tourists and residents, to the downtown

area. SWA decided to create a public place that would be more useable by pedestrians, a place that would attract people to cultural and civic events both day and night. This plan was set up to draw people to the downtown area first, thus attracting addtional redevelopment to the area from private sources.

The SWA Plan for Lewis Avenue Corridor was arrived at with much input from the community. It creates a new central spine through the heart of downtown Las Vegas connecting the Office Core District, including the Clark County Detention Center at one end with the new Federal Courthouse Building at the other end three blocks away. Along the way there are historic buildings, City, County and Regional administrative buildings and connections to the Arts District and residential neighborhoods nearby.

Before development, the Lewis Avenue environment was brutal. The Buildings were generally lacking in any

detail that could provide a welcoming human environment. The buildings had no cushion around them, but were jammed right up to the traffic-dominated roadway. There were grossly undersized sidewalks that were completely devoid of shade. The environment was not inviting, in fact, pedestrians were rarely seen on the sidewalks. SWA began by reallocating the space within the street corridor. Two of the four motor-vehicle travel lanes were removed. The resulting "found" space was given back to the pedestrian, which resulted in 20' wide sidewalk zones.

The linear street space linked the public spaces previously created by SWA at the Federal Courthouse site and at the Regional Justice Center. The Courthouse, sited 10 feet above street grade, was planned to include a fountain feature at its front door. A design concept emerged from this opportunity. The fountain would become the "source" for a "wash" that would meander along the street and would become the backbone of the streetscape environment.

The design process began with a Community Workshop to which the public was invited and in which they were encouraged to participate. The Las Vegas Arts Commission and local artists stepped forward to participate and, as a result, the plaza evolved as a public art venue. And now that the sidewalks would be increased in size, a double row of Ash Trees could line both sides of the street and would create dappled shade as well as a human scale and a buffer against unfriendly architecture. Groves of palm trees were planned to be planted in such a way as to interrupt the ash tree canopy, announcing the primary entries to the Regional Justice Center and the mixed-use center.

Because the street trees would be of such great importance, SWA designed a continuous planting trough filled with structural soil so that the trees could more closely reach their potential height and spread dimensions and have greater longevity. Smaller, desert trees including the Palo Verde and Mesquite were used along the wash as specimens and groupings. The central wash, an artistic interpretation of a natural system, is a meandering cobble-lined feature that is recessed two steps below sidewalk grade so that people can sit along it's edges. The exposed edge also reveals the color and texture of the natural landscape. The wash changes character as it moves to each successive block and evolves from an organic form to a system of rivulets. Fountain features along the way serve as water sources and a cascading water wall merges out of the lower basin and is oriented along the plaza axis and perpendicular to Fourth Street to announce the space to those headed for the Fremont Experience. Native, arroyo plant materials are integrated into the wash to allude to the natural desert landscape. Pedestrian crossings are provided periodically by means of either a warm-toned concrete bridge or a more informal grouping of natural flagstones.

1 刘易斯大道线形水景区的一端
2 水景

Project materials and details are simple and of a hearty scale to reference both the urban condition as well as to minimize on-going maintenance requirements. The street furniture, consisting of benches and drinking fountains, expand the warm-toned concrete materials palette and reinforces the objective of low-maintenance components. SWA chose to use the existing downtown streetlight standard, a simple round shoebox fixture, so that meager project funds could be directed into more important components.

A distinct design challenge came from the City's standard storm drainage design approach. Due to the limited rainfall in Las Vegas the City did not typically implement a storm drainage system but simply allowed the streets to flood during isolated rainfall events. To alleviate the problems that would result for the pedestrian in such an environment, SWA tipped the entire ground plane to a collection point that rids the street of standing water and further supports the goal of pedestrian use.

The artists' participation yielded a special detail for the street. Dubbed "Poets Park" by the group, 18 artists created poetry and prose that was subsequently etched into the pedestrian bridge paving and has become an integral part of the street's identity. Lewis Street Corridor's impact on downtown redevelopment is already very apparent. It is a very successful prototype for getting people out of their cars and onto the sidewalks. The Lewis Street Corridor, through a collaborative public and private partnership, has created a lively civic space for those who call Las Vegas their home.

1 建筑物入口处的袖珍公园
2 刘易斯大道上的公共雕塑
3 线形街道

查普特佩克公园喷泉步道

Fountain Promenade at Chapultepec Park

撰文：Jimena Martignoni　　　图片提供：Francisco Gomez Sosa　　　翻译：李沐菲

查普特佩克公园位于墨西哥城，是美洲最大的公园。该公园曾是阿兹特克人的圣地，他们在此修建道路、种植树木并开渠引水，用以供应他们的首都特诺奇特兰。因此，这可能也是一座最古老的公园。1530 年，西班牙皇室将公园的所有权返还给当地居民，公园经过重新设计在 1906 年 ~ 1907 年间重新开放。该公园包括国家最重要的博物馆、剧院、舞蹈学校、动物园以及一大片湖区。

然而，随着时间的流逝，这里出现了大批的路边摊贩、缺乏整体管理及定期维护，致使该公园的整体形象大打折扣。鉴于公园这种混乱的局面，在 21 世纪初期，一个由当地居民和商人组成的团体——林地市民委员会及改造委员会共同筹资，对公园进行全面升级改造。委员会聘请了 GDU 设计公司，对公园进行整体的升级改造。总体规划由 GDU 设计公司提出，最初是为满足墨西哥政府进行城市研究而制定的。在对园区内的树木进行了彻底详尽的调查后，设计师最终施行了三项措施：基础设施建设、树木的移除及修剪以及路边摊位的重新安置。为此，设计师在园内共划分出 15 个区域，虽然环境特征相似，但以上三项措施均会因地制宜地得到贯彻执行。

作为"基础设施"的一部分，喷泉步道位于该公园的北部，将墨西哥国家人类博物馆和 Tamayo 当代艺术博物馆连接起来，是总体规划中的关键设计元素之一。考虑到人行道的轴线穿过查普特佩克公园内游人最多的区域，这条 250m 长的喷泉步道被设计成了可供游人放松、约会及休憩的绝佳地点。

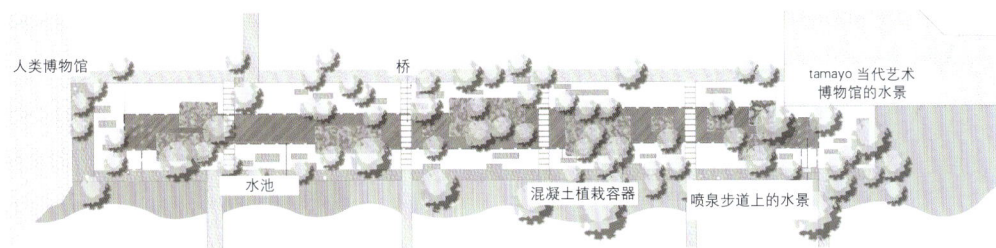

平面图

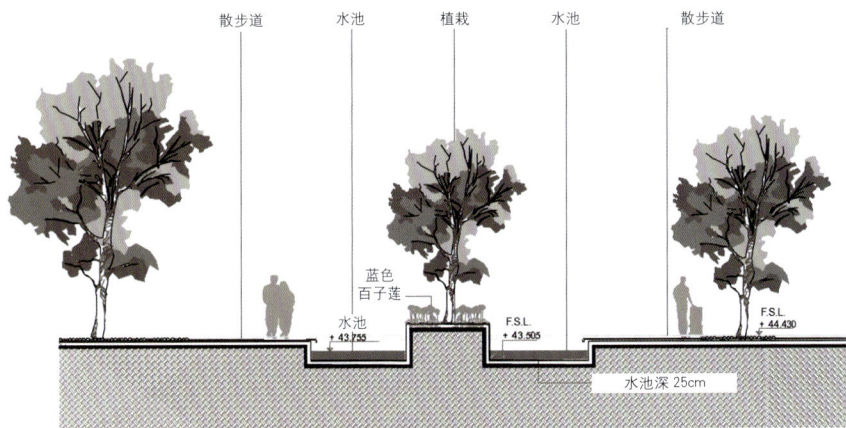

剖面图

1　公园中的喷泉极具动感
2　蓝色百子莲种植在水景边缘，
　　人们在设计新颖的坐椅上休憩
3　经维护和设计后公园的自然景象

　　过去，该地区没有一条合适的人行步道来连接两座博物馆，设计师及时发现了这一点并立即填补了这一空缺。为了建造这条人行步道，设计师必须对部分区域进行适当的改造，这也是总体规划中的主要目标之一。"树木的移除及修剪"实际上是一项比较有争议的措施，设计师与委员会的成员通过组织和参加一系列公共会议来解释这一措施的实施理由以及方式。在认真且详尽的计划指导之下，公园内的植被向着更加健康的态势发展。

　　喷泉点缀在这片郁郁葱葱的林木之中，两侧的视觉效果极佳、景色迷人，这里不但为树木提供充分的生长空间，还可以使阳光照射进来，营造出美妙的光影效果。过去，树木曾过度生长、十分密集，犹如厚厚的遮光器；如今，树木之间充足的空间不但可以使其接受足够的阳光照射，草坪也因此可以更快、更充分地生长，使地面看起来更加绿意盎然。

　　喷泉宽 20m，在步道的两侧呈线性分布。循环水、植被以及路面的铺装等元素构成了一幅极具动态魅力的景象，成为了公园里的焦点。水景元素以多种形式展现出来，大部分的喷泉水面是平静的，一些水柱点缀其间，有些喷泉之间还铺有石板，水在石板之间流淌，像小瀑布一样潺潺流向下一层。

　　与喷泉对应的区域是一片植被，与水景形成鲜明的对比。蓝色与白色的百子莲以及地被植物如同铺在地面上的地毯一样，这也是树木被移除修剪之后使得灌木和草坪都能够充分生长的结果。

　　为了方便游客休息及观景，景观设计师采用了一些定制的设施和新型的照明设备。根据人体工程学原理设计的色彩明亮的长椅无序地摆放在喷泉的两侧，营造出一种欢愉而又轻松的水边氛围——人们安坐于此，休息片刻，环顾四周，感受着连绵不绝的水声带来的惬意心情。

　　在这个喧嚣的城市中，喷泉步道俨然已成为一个小小的世外桃源。

Bosque de Chapultepec is a 686-hectare public park, the largest in America, located in Mexico City. It's also probably the oldest one, for it used to be a sacred area for the Aztecs, who built roads, planted trees and channelled the water that supplied their capital, Tenochtitlán. Ceded in 1530 to Mexico City by royal cedula for the use of its inhabitants and subsequently redesigned and reopened between 1906 and 1907, this park presently houses the country's most important museums, theatres, a dance school, zoo and an area with large lakes.

Over time, however, the unplanned growth of the areas with street vendors, the lack of integral interventions and of a regulated maintenance plan led to a very neglected general image of the park. This chaotic situation was what, at the beginning of the 21st century, motivated a group of citizens and local businessmen to get together and raise the funds to elaborate a master plan of rehabilitation. The group, called Consejo Rector Ciudadano de Bosque (Civic Board of Citizens for the Woodland) and the Fideicomiso Pro-Bosque (Pro-Woodland Board of Donors) hired Grupo de Diseño Urbano (GDU) to radically change the state of the park.The Master Plan proposed by Grupo de Diseño Urbano was initially based on an analysis requested by the Mexican Government to the City's Studies Program of the Universidad Nacional Autónoma (2001-2002). With the addition of an exhaustive study of the phytosanitary state of all trees within the park—made by the Universidad Autónoma Chapingo and the Universidad Autónoma Metropolitana—the designers finally determined three particular courses of action: "Infrastructure" "Tree removal or pruning" and "Reorganisation of the street vendors". These actions were implemented according to a physical differentiation between fifteen zones within the park, each of which had similar environmental characteristics.

As part of the "infrastructure" course of action, the fountain promenade that connects the National Museum of Anthropology with the Tamayo Contemporary Art Museum, in the North area of the park, is one of the key design elements of the Master Plan. Thought of as a pedestrian axis through one of the most visited areas of Chapultepec Park, this 250 meter-long fountain has became a place of relaxation, encounter and shelter.

Formerly, the site lacked a proper pedestrian connection

between the areas of the two museums and the designers found that this was an undoubtedly necessary addition. In order to create this connection they had to "clean up" part of the area, a task that was also one of the main objectives to be met by the Master Plan. The "pruning and removal of trees" course of action was actually quite controversial and the designers, together with the members of the Board of Citizens that promoted the rehabilitation plan, had to organize and attend a number of public meetings during which they had to faithfully explain the reasons and methods behind this decision. Carried out with a very responsible and thoroughly planned system, the cleaning up process led to a generally much healthier planting's state.

The fountain travels through green masses of trees which now look vigorous, but even more important and more visually verifiable is the present image of the two sides of the fountain where trees now have room to grow and where the sunlight comes through, creating beautiful effects. Before, overgrown trees had shaped a thick formation, almost impossible to pass, which acted as a light blocker; today, with more free space between trees and consequently more possibilities of light, even the ground looks greener because the lawn grows stronger and faster.

The fountain itself, which is 20 meters wide, is laid out as a geometrical linear piece framed on each side by respective paved pedestrian paths. Alternating water, planted and paved surfaces, this element turns into a very dynamic and attractive focal point in the park. Water appears in various different manners. Surfaces of still water cover most of the fountain's area while a few spots are dotted with water jets; at those points where the fountain is interrupted by stepping stones (offered as a transversal connecting pedestrian element), water runs between them flowing on, as subtle falls, into the next water surface.

Alternating with the water areas and symmetrically placed along the full extent of the fountain, fully planted surfaces which reminiscent of a natural patchwork complete the design. Homogeneous carpets of blue and white agapanthus and carpets of groundcovers create a "sotobosque", or bottom green surface, which profusely grow underneath the green canopy generated by the now restored tree tops.

In order to provide resting and watching possibilities for visitors, the landscape architects incorporated custom-designed furniture and new lighting. The colourful and ergonomic benches irregularly positioned at both sides of the fountain help defining a joyous yet relaxing ambiance by the water; people sit, have a rest and watch others while being soothed by the constant babbling water.

Within the largest park of one of the largest and most chaotic cities of the world, this fountain promenade embodies, indeed, an exceptional spot of respite.

变幻的空间体验 —— "螺旋"桥

Changing Experience of Space — The Twist

撰文：West 8 城市设计 & 景观设计　　图片提供：Jeroen Musch　　翻译：高明

由 West 8 设计的新的自行车道——螺旋桥，被放置到其最终的目的地——荷兰弗拉尔丁恩市 Vlaardingse Vaart 河岸边。该桥的运输以及精确到毫米的嵌入桩基和桥墩的过程令人印象深刻，吸引了数百名围观者。该桥于 2009 年 4 月正式开放，是荷兰南部地区自行车道网络的一个关键连接点，也形成了与 Broek 圩田之间的城市化及休闲路线的联系。这种新的联系需要在这一地区有一处鲜明的标志物。

该桥位于城市边缘的一片天然绿色环境中，景色十分吸引人，令行人折服。在不断发展的城市中，自然景观很稀少。这座相对孤立却引人注目的桥能够在跨越河流的同时捕捉其所在地景观的美感，并且提升了景观的质量。从不远处的高速公路望去，当小船顺着种植着高大绿色植被的河道从桥下驶过时，这座桥又可以被看做是 Broek 圩田的一个入口。

该桥预先在一个临时的特制集装箱里建成，由 400 根钢管焊接而成，经过镀锌、漆成红色，形成了一个独特而富有动态的结构。剖面为正方形的三维钢质构架呈螺旋状绕着桥的横轴旋转，这一旋转的空间框架结构以其持续变化的视觉体验吸引行人的注意。入口处的正方形框架在桥的中部逐渐演变成了有两个顶点的钻石形状。这个通透的钢质构筑物成为了人行道和自行车道网络中新的休息和约会的场所。当骑自行车穿过桥时，这些不断变化的剖面增强了该桥看似在进行旋转的效果。

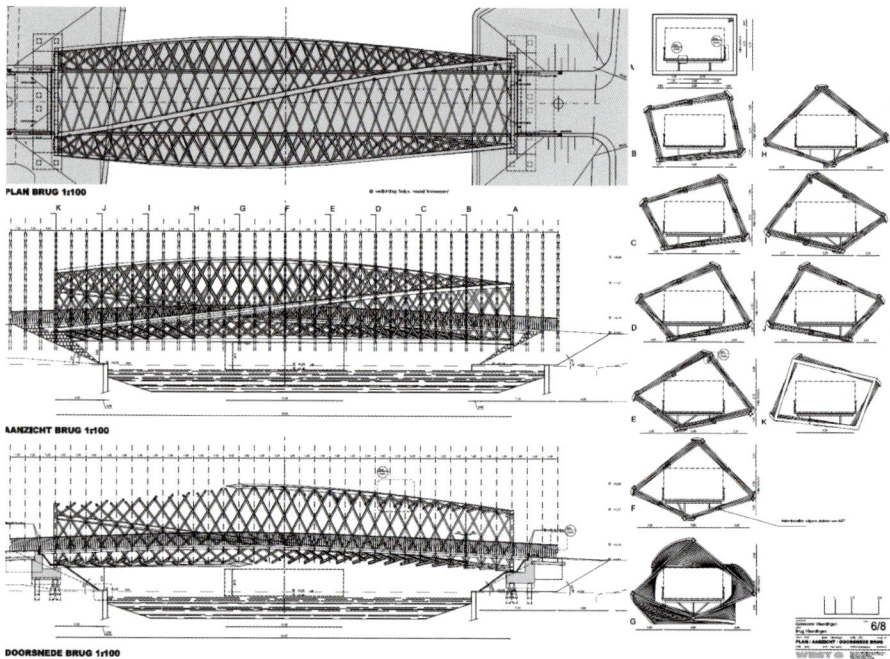

PLAN BRUG 1:100

AANZICHT BRUG 1:100

DOORSNEDE BRUG 1:100

6/8

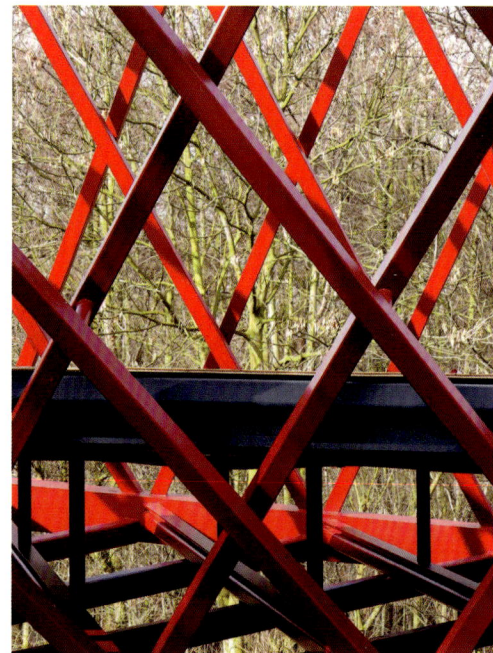

The new pedestrian bicycle bridge designed by West 8 for the municipality of Vlaardingen, was officially opened in April 2009 to allow pedestrians to cross the Vlaardingse Vaart. The bridge in Vlaardingen is a crucial link within the regional bicycle path network of the province of South Holland. It also forms an urban and recreational route connection with the Broek polder. This new connection calls for a strong identity in the area.

The bridge was prefabricated on site in a specially erected temporary shed of stacked containers. Its construction comprises 400 steel tubes that were welded together and then galvanized and painted red to create a unique and dynamic structure. The three-dimensional truss of the bridge twists on its horizontal axis. By virtue of its location in a natural green setting on the edge of the city, the bridge serves to tempt and surprise its users. The natural character of the setting is scarce within the bounds of the continually developing city. An expressive sculpture with a withdrawn character establishes the moment of "crossing" and so doing captures the qualities of the immediate landscape. From the nearby highway the bridge can be identified as a point of entry for the Broek polder as small boats sail underneath and along the canal with tall greenery on both sides stretching out over the horizon.

The three-dimensional truss of the bridge, made from square steel profiles, twists on its horizontal axis. This twisted space frame construction absorbs the user in a continually changing experience of perspective. The rectangular frame at the bridge entrance evolves into a double-height diamond form in the middle. A transparent cathedral of steel becomes a new rest and meeting place along the footpaths and bicycle network. The continually changing sections reinforce the twisting movement as a bicyclist moves through the bridge.

光之步行道
Promenade of Light

撰文：Tonkin Liu　　图片提供：Keith Collie　　翻译：王玲

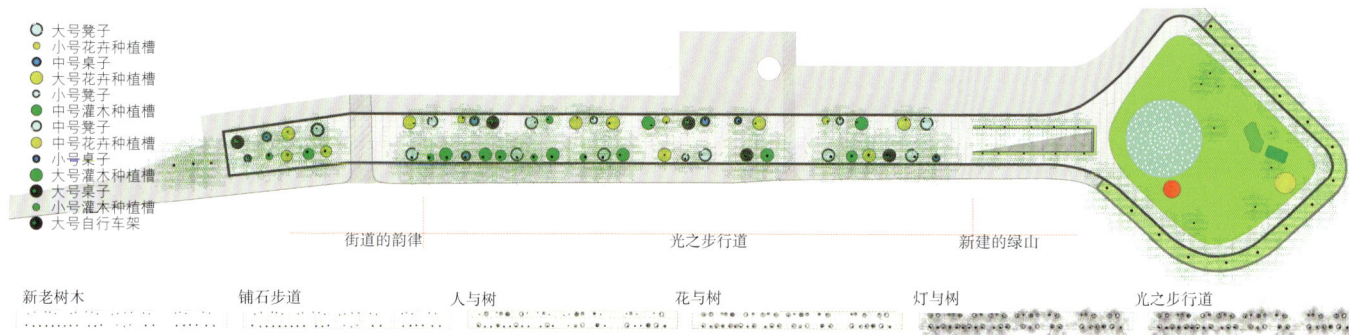

大号凳子
小号花卉种植槽
中号桌子
大号花卉种植槽
小号凳子
中号灌木种植槽
中号凳子
中号花卉种植槽
小号桌子
大号灌木种植槽
大号桌子
小号灌木种植槽
大号自行车架

街道的韵律　　　　光之步行道　　　　新建的绿山

新老树木　　铺石步道　　人与树　　花与树　　灯与树　　光之步行道

该项目是 Tonkin Liu 公司在 2004 年赢得的一项设计竞标。设计师对伦敦老街以西的人行道区进行改造，将伦敦金融区入口那块被忽视的地块彻底地改头换面，使其更加安全便捷、美观实用。该项目堪称是将创新高效的设计融入城市环境中的成功典范。

该项目竞标由 EC1 社区新政、伊斯灵顿区和伦敦交通局共同发起，他们希望改善位于老街地铁站和巴斯街之间的老街北侧的公共空间。该竞标的内容包括草坪和商店前环岛西侧的道路改造建议以及扩宽周围环境的战略规划。

项目改造的主要目标包括三个方面：首先，重新规划步行人流密集的伦敦中部地区，令交通更加便捷；其次，将一个利用率低的绿色空间转变成坐席充足的高品质公共空间；再次，提高场地照明和便捷性的设计，营造出更加安全、舒适的环境。

场地上一块利用率低的草坪或将行人引向熙攘的街道，或引向商店前狭窄的路面上。设计团队发现场地的特点是适于不同的群体使用，这些使用者包括通勤者、购物者、物流人员、学生、老人、上班族或者一家人——他们在不同的时间以不同的方式使用着空

间，最主要的是他们以不同的节奏使用着空间。因此，设计师力图打造一处更加舒适宜人的公共空间，不仅能够满足使用者的不同需求，而且还能在白天吸引更多的人至此，在夜晚则可以充当安全舒适的交通动线。

Tonkin Liu 公司的设计改变了人们欣赏街景的步伐。他采用绿树成荫的传统的步道方式，重新安排的设计元素不仅吸引人们纷沓至此，更丰富了人们的身心体验，促进了人们的交流互动。

场地上 21 株成熟的悬铃木为新规划提供了基础，确保了最低限度地破坏原有植被。原有树木和 18 株新栽植树木分成两排，形成一条独特的林阴路。中心草坪和低矮的边界墙被改造成一条石块铺成的中央步道，任何人都可以使用这条步道，包括老年人和行动不便者。占地 2015 ㎡ 的步道在增强场地硬质景观效果的同时，也在高峰时间承载了更大的人流量。

在规划中，每一棵树的周围都安装一个圆形种植槽，这些种植槽可以充当桌椅、花盆或循环设施。圆形的景观主题在道路铺装上也被不断地重复使用，形成一簇簇的圆形图案，将人们聚集在一起，为生活节奏不同的使用者营造出凉爽宜人的城市休闲空间。

白天，阳光透过树叶的缝隙在浅色的铺石路面上洒下斑驳的树影；夜晚，路面仿佛一块画布，各种灯光在上面自由生动地渲染开来。

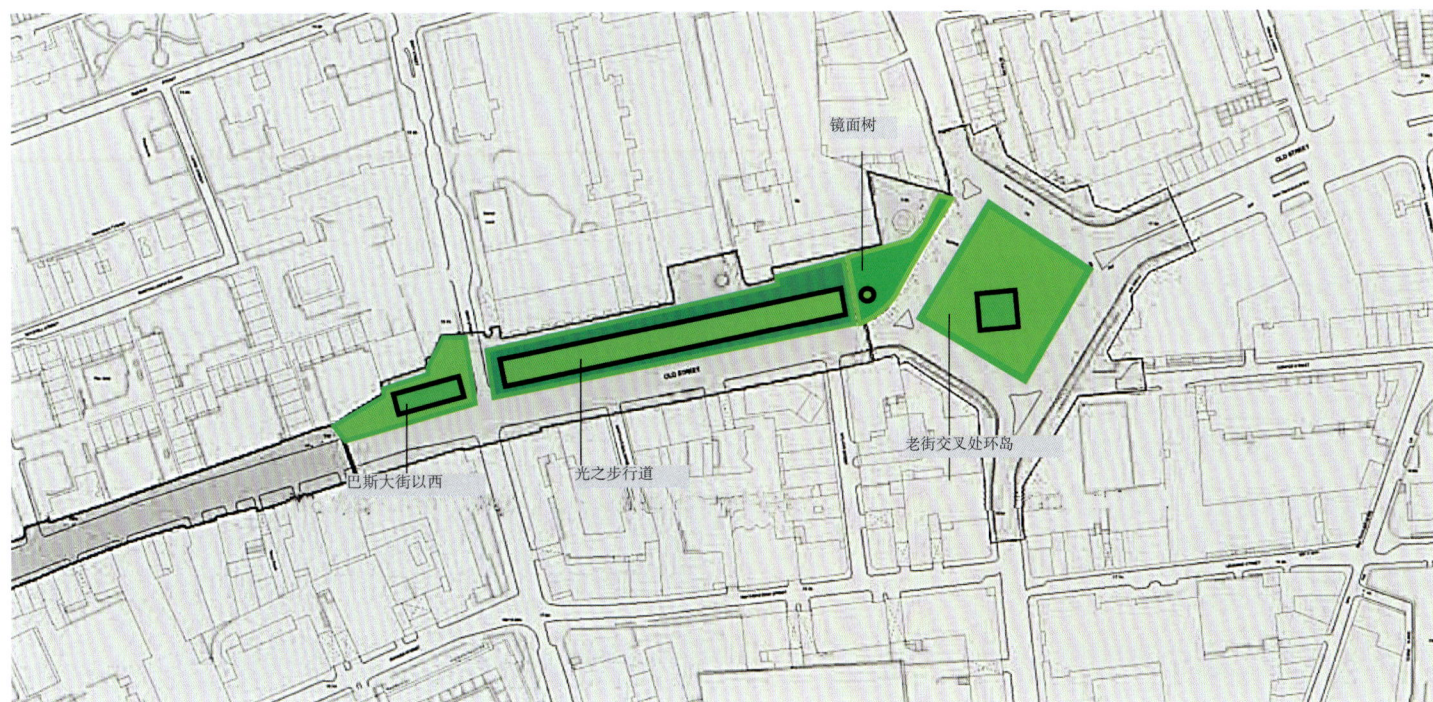

步道的照明设计是使空间在夜晚更加安全、美观的一个主要考虑因素，也是规划中的一大亮点。灯柱上的多向聚光灯既可作为基本的步道照明，也可在照射到植物时产生剧场般的效果。阳光跟踪时控开关在不同的时间会自动调整设置，如周末和平时、夏季和冬季，还能够在光照强时存储一定的能量。

项目影响

议会通过对该项目开始前和结束后的街头调查显示出设计对使用者的影响。行人普遍反映步道比以前更加安全，认为改善后的街景能惠及当地企业。使用者对空间满意程度也从之前的 59% 提升至 80%。

此外，该项目还赢得诸多的建筑奖项，如 2007 年英国皇家建筑师协会伦敦篇章奖、2007 年首项优秀建筑奖入围项目、2007 年英国建筑工业奖、2007 年再生项目奖、2008 年照明设计奖、2008 年天然石材奖和 2008 年第五届城市公共空间欧洲奖。

总体规划

项目的最新阶段——被抬起的镜子环绕着一株玉兰树，于 2009 年春季揭开了神秘的面纱，它为步道画上了圆满的终止符，同时也是地铁站的入口。Tonkin Liu 公司将在项目的下一阶段着手进行老街交叉处环岛、老街车站的大厅和入口的设计，他还将与伦敦交通局一起开展该区的城市设计架构。遵从伦敦交通局的交通影响和成本评估的要求，设计师将对整个场地及其交通进行重新评估，为行人、骑自行车者和园艺工作者重新改造原来的交通岛。

Promenade of Light, Tonkin Liu's 2004 competition-winning scheme for the refurbishment of the pedestrianised zone west of Old Street London, is a superb example of innovative and effective design in the urban realm. It has completely transformed a neglected area of the city at the gateway of London's financial district, making it safer, fully accessible, better looking and easier to maintain.

The project began as an EC1 New Deal for Com-munities, London Borough of Islington and Transport for London initiative to upgrade the public space on the North side of Old Street between Old Street underground station and Bath Street. A high profile architecture competition was launched inviting strategic proposals for the redevelopment of the grassed area and pathway West of the roundabout in front of a parade of shops, as well as a wider strategy for the surrounding area.

The principal aims of the project were threefold; to regenerate an area of central London prone to high volume of pedestrian traffic, making it more accessible; to convert an underused green space into a high quality public space with ample seating opportunities; to improve lighting and accessibility, making the site a safer, more pleasant environment.

The site as found was characterised by an underused grassed area, which orientated pedestrians along the busy road or on a narrow area of paving in front of the parade of shops. Winners of the design competition, Tonkin Liu, observed that the site was characterised by a range of user groups—commuters, shoppers, couriers, school children, elderly, office workers, families—and that these user groups inhabit the space in different ways, at different hours of the day and most importantly, at different paces. The intention

was to create a more inviting public space, which would appeal to all users, invite occupation in the day and provide a safe and pleasurable route in the evening.

Tonkin Liu's scheme has altered the pace of the existing streetscape with the introduction of a traditional tree-lined promenade, an arrangement which invites dwelling and celebrates the experience of walking and interaction, allowing ideas to unfold and relationships to grow.

The 21 mature Plane trees on site provided a basis for the new scheme, ensuring minimal disruption to the existing planting. The existing trees were supplemented with an additional 18 trees to delineate two distinct rows. The central grassed area and boundary low wall have been removed and replaced with a central stone promenade which is fully accessible to all, including the elderly and mobility impaired. Measuring 2015 sqm, the promenade increases the hard

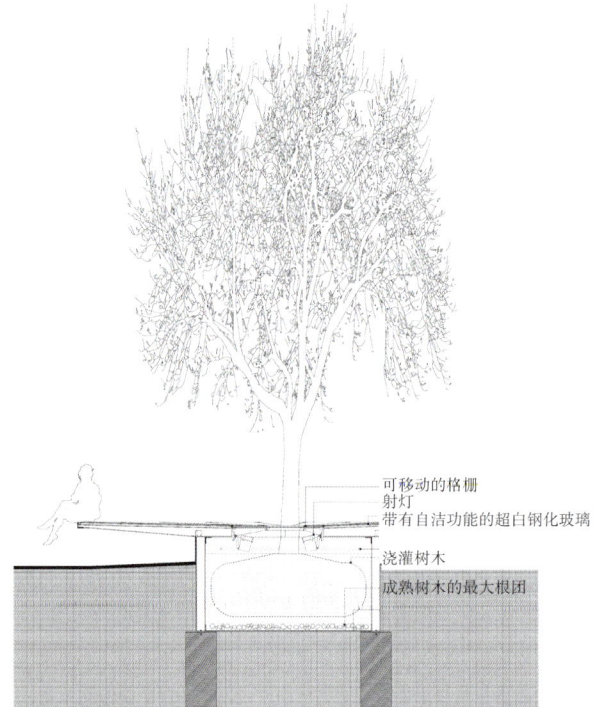

可移动的格栅
射灯
带有自洁功能的超白钢化玻璃

浇灌树木
成熟树木的最大根团

landscaping on the site, accommodating a greater flow of pedestrians during rush hour.

Every tree in the scheme has been framed by a ring, which takes the form of a bench, a table, a planter or cycle facilities. The ring motif is repeated in the paving and creates an arrangement of clusters of rings to bring people together and create urban rooms under the shelter of the trees and clearings to accommodate the varying rhythms and paces of the broad demographic of users.

For the promenade a light stone was specified to maximise the dappling effect of sunlight and shadows through the canopy of trees during the day and to provide a canvas for more dramatic light projections at night.

Illumination of the promenade has been a principal consideration in making the space safer and more attractive at night. The scheme makes a feature of the lighting, introducing lamp columns with multi-directional spotlights, which combine basic footway illumination and more theatrical lighting effects such as the illumination of the tree canopy and planting. A suntracking timed switch has been integrated and programmed to vary the configuration between weekends and weekdays, summer and winter and to conserve energy in bright periods.

Project Successes

Pre and post implementation on-street attitudinal surveys by the council indicate how the design has affected users. When questions pedestrians said they felt safer on the promenade, recognising the benefit of the improved streetscape to local businesses. When asked if users enjoyed being in the space, 80% of interviewees strongly agreed in comparison to 59% prior to the implementation of the new promenade.

The project has also been recognised in architectural awards; it won a RIBA London Award in 2007 and was a finalist in the Prime Minister's Better Building Award 2007, British Construction Industry Awards 2007, Regeneration Awards 2007, Lighting Design Awards 2008, Natural Stone Awards 2008 and the 5th European Prize for Urban Public Space 2008.

Masterplan

The latest phase of the project—a planted magnolia tree set into a lifted mirror surround—was unveiled in Spring 2009 to mark the terminus of the promenade and the entrance to the underground station. In the next phase Tonkin Liu will address Old Street roundabout, Old Street station concourse and entrance. Tonkin Liu is working with Transport for London to implement an Urban Design Framework for the area. Subject to Transport for London's assessment of traffic impacts and cost, the whole site and traffic is being reappraised, to reclaim the existing traffic island for pedestrians, cyclists, and landscaping.

布宜诺斯艾利斯城的"行人优先"计划

Priority for Pedestrians in Buenos Aires City

撰文：Jimena Martignoni　　图片提供：Design team Buenos Aires　　翻译：申为军

大街重修总体规划图

阿根廷首都布宜诺斯艾利斯的建筑和城市规划向来带有明显的欧洲风格——宽阔的景观大道和街巷、各种大型公园和露天广场、市民休闲场所。然而，在过去的30年间，由于城市的迅猛发展，布宜诺斯艾利斯也出现了很多问题，如交通拥堵、社会不公正、缺乏步行空间，这种典型的拉美模式已成为城市的痼疾。

自21世纪初，市政府一直在制定环境美化的规划。2003年，针对尤其混乱拥堵的市中心，市政府明确了几项任务，其中最重要的一个目标就是减少小汽车流量，提高公共交通的能力。实现这一目标显然需要一个长期的过程，市政府首先集中精力改善城市的某些区域，并致力于实施"行人优先"措施。其中包括以下几项改造项目：修缮七九大道、把里斯·萨克特斯街和雷孔基斯塔街改造为步行街、改造 Bullrich 大街，这四个不同的项目在功能和效果上是彼此关联的。

七九大道的修缮

七九大道有100m 宽，号称世界上最宽的街道之一，是贯穿布宜诺斯艾利斯城市南北的一条主干道，两侧分布着若干座阿根廷最具标志性的社会机构建筑。七九大道始建于1937年，20世纪70年代中期建成最后一段街道。几十年过去了，主路两侧的林阴大道种植的很多当地树木如今已经郁郁葱葱，绿阴如盖。七九大道的主路宽约60米，共有14条车道，南北双向各半，中间是一道狭窄的、安全性差的隔离带，供行人穿越马路时停留。此外，两条林阴大道与街边的建筑之间还隔着一条街，各有四条车道。随着时间的推移，林阴大道及其铺装部分急需修缮，主路的中央隔离带也有待改善。

帕萨赫·里斯·萨克特斯
400/500

帕萨赫·里斯·萨克特斯
300/400

里斯·萨克特斯大街总设计图

Bullrich 大街整体规划图

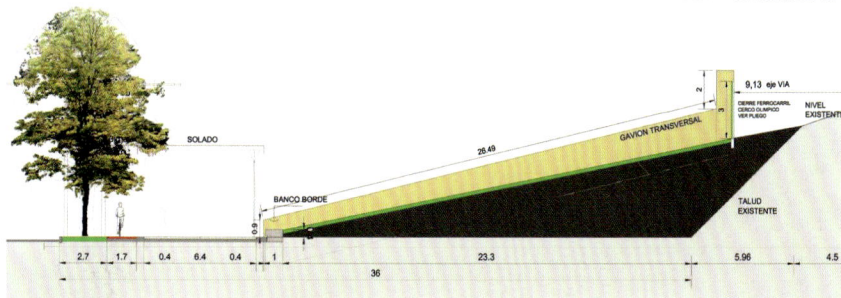

Bullrich 大街剖面图

| VEREDA | AV. BULLRICH | EX-PLAYA REGULACION COLECTIVOS / EX-DEPOSITO POLICIA | TALUD EXISTENTE / VIAS FFCC |

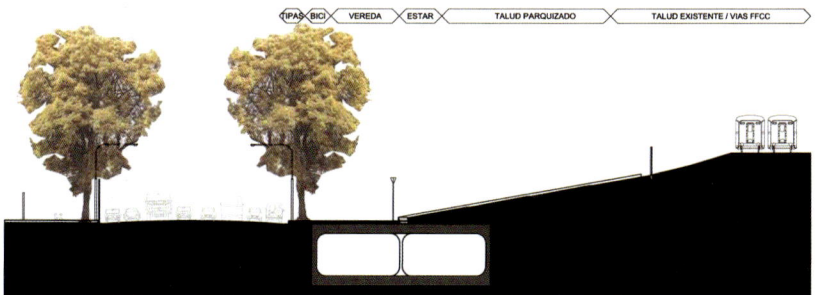

| TIPA BICI | VEREDA | ESTAR | TALUD PARQUIZADO | TALUD EXISTENTE / VIAS FFCC |

| VEREDA | AV. BULLRICH | PARQUE LINEAL BULLRICH | TALUD EXISTENTE / VIAS FFCC |

Bullrich 大街剖面对照图

修缮工程分为四个阶段，目前已经完成了前两个阶段，将实现如下三大目标：改变单纯以小型机动车为导向的城市中轴线设计，营造更为便捷的出行环境，重新评估那些树龄较大的当地树种的价值。第一阶段的任务是把两条林阴大道改造成线性广场，拓宽中央隔离带。

新建广场的地面铺装以经济且坚固耐用的混凝土和花岗岩碎片为主，与相邻的广场在视觉上形成连续性。中央铺装区域添置了新的坐椅和灯柱，边缘种植草本植物和低养护的灌木品种——在原有树木的高大树冠的映衬下，如绿色的地毯一般延伸出去。如果这些植物因为季节的原因枯死，设计师将会补栽相同的品种。

每个林阴大道广场都根据其原有的状况进行设计，有的装有老式护栏，有的配置种植槽，有的通往地铁或停车场。通往地下公共停车场的小型机动通道原本标示不清，容易使人迷惑，现在则改造成 U 形通道以绿色示人，而行人的等待区仍被优先对待。

新的中央隔离带由先前的 1.2m 拓宽到 6m，有些地段则达到 8m。设计师以几何形式种植各类高度不同的灌木，其间斜穿着窄窄的小路。这些小路既是养护通道，又形成了充满动感的空间，视觉效果尤其显著。

开辟步行街

作为"行人优先"计划的一部分，市政府决定将市中心某些街道或其中一部分改造成步行通道和步行区。这项任务并不轻松，起初，有些项目具有很大的争议性，遭到一些人士（主要是机动车司机）的强烈反对。正是基于这个原因，改造项目采取分阶段完成

1 重修后的里斯·萨克特斯大街下段

的方式，在街区的选择上也相当谨慎。无论如何，人们普遍支持建立新的步行街，整个计划也得以完整实现。

被改造为步行区的首批街道中包括里斯·萨克特斯街。这条街很短，只有两个街区，长 200m，原本被作为一条封闭街道或通道使用，车流量很低。改造后的这条街作为紧急情况下的步行道和住户停车时的出入通道。只在路面中央留下一条原有的鹅卵石铺装作为装饰，两侧则使用混凝土和花岗岩碎片。设计中还新增了隔离墩、线形下水道、行道树和坐椅，现在的街景是如此迷人，很难想像它就靠近布宜诺斯艾利斯最繁忙的干道和大街。

雷孔基斯塔街与里斯·萨克特斯街垂直相接，也被选定改造为步行街。这条街道很长，改造工程相对复杂，因此计划分三个阶段完成，每个阶段改造四个街区，约 400 米长，第一阶段的改造目前已经完成。

雷孔基斯塔街被改造为步行街的主要原因是整条大街上分布着众多的酒吧和餐馆，已经成为人气很旺的新兴街区。改造后，人行道被拓宽并与原有的街道高度上一致，以隔离墩相隔，还种植了开花的树木。

如今，这些酒吧的户外区与改造后的充满生机的步行区融合在一起，商业区的上班族在午饭时间和下班后常常在这里休息，其间还夹杂着那些无时不在的旅游者。

Bullrich 大街改造

这条大街总长约 800 米，位于公园和景观大道的密布区，通往诸多重要的繁华街道，旁边是一条高架铁路，过去一直供公共汽车停靠和落客使用。

由于落客区环境不佳和缺乏养护，市政府决定对这条大街进行全面改造，这也是城市改造计划中的一

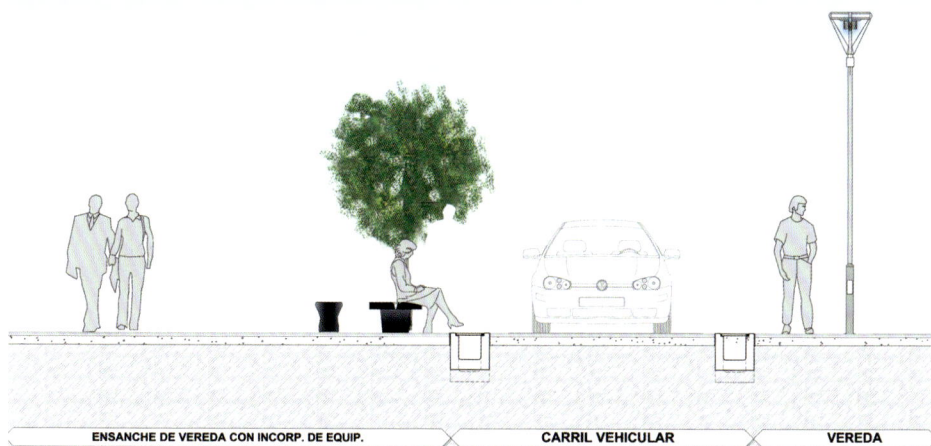

ENSANCHE DE VEREDA CON INCORP. DE EQUIP.   CARRIL VEHICULAR   VEREDA

雷孔基斯塔大街一角

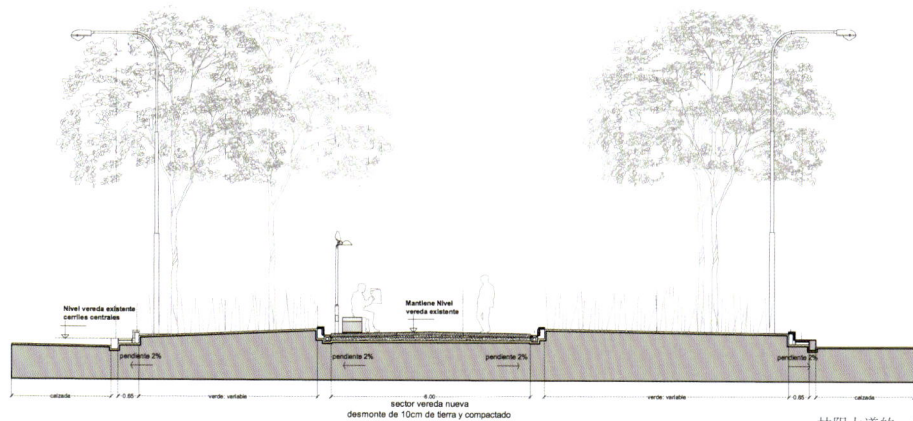

林阴大道的一角

部分。改造目标是把这里变成一块城市绿洲，供人们散步休憩，并起到连接以汽车为导向的市中心各区域的作用，因此在设计上强调简洁、优美的景观和街景规划。

高架铁路和人行道的高度落差约有 7 米，为了解决这一问题，在二者之间修筑了陡峭、坚固的斜坡，上面自然生长着一簇簇的植物，斜坡旁就是铺装单调的公交停车区。这里有城市中最早种植的人工林，沿街种植着一排高大的当地树木，穿过 Bullrich 大街对面也有一排大树，提供了良好的封闭性和垂直绿化。

基于以上现状，设计师决定在场地原貌和新增构建之间创建空间平衡，勾勒出这块区域的纵向轮廓，并利用现有路堤形成更为水平的地面。设计师采用一系列直线式种植组合模式，包括各种不同品种，如人行道边种植地被植物、低矮而艳丽的球茎植物、草坪，铁路护栏边则种植相对高些的观赏草。在这条绵延的植被带上，每隔 42m 就被一道里面装满石块的丝网矮墙分割开，形成连续的韵律，并削减了由于植株高低不同及其所需表层土深浅不一所形成的轻微高度差的问题。

这些植被带覆盖了原有的路堤，一度陡峭的地势得到了缓和，斜率由 60% 下降到 20%，并一直延伸到附近的公交落客区。曾经被当做落客区使用的铺装区与现有的人行道连接起来，高度也达到一致，形成了颇为宽敞的步行空间——一侧是生长了多年的大树，另一侧则是绿意葱葱的舒缓斜坡。

步行区安装了新的灯柱，布置了造型奇特的水泥坐椅，其原始设计可追溯到上世纪 70 年代，此次专门为这块空间和市中心的其他公共空间而重新制作出来。

这些城市中心地块的改造十分成功，结果也相当令人满意，该项目探讨并证明了城市中的人们对绿色步行空间的巨大需求。

1

Buenos Aires, capital city of Argentina, has historically been recognized by the European style of its architecture and urban planning: wide landscaped avenues and streets, large parks and intimate plazas, and places to be enjoyed by people. However, during the last three decades the city has been growing at quite a fast pace and the Latin American model, mostly characterized by major traffic problems, social inequity and lack of pedestrian areas, has been taking root.

Since the turn of the century, the city government has been working on the creation of a better environmental plan and, in 2003, delineated some specific tasks for the central area of the city, which was particularly deteriorated. Among the most important objectives for this area were the reduction of cars and the improvement of public transportation; although this last one would only be realistically possible to complete in the long term, the city government focused on the renovation of certain pieces of the city and on a program that was called "Priority for pedestrians" or Prioridad Peaton.As part of those, four different projects remain relevant, for both their functional and formal results: the renovation of 9 de Julio Avenue, the pedestrianisation of Tres Sargentos and Reconquista Street and the transformation of Bullrich Avenue.

Renovation of 9 de Julio Avenue

9 de Julio Avenue is one of the widest avenues in the world, at more than 100 meters, and the main north-south traffic line of the city; framed by some of the most emblematic institutional buildings, it was first built in 1937 and its last segments were added in the mid 1970s. Two boulevards, planted with native trees which, after decades, look large and luxuriant, run along the avenue framing a central segment of approximately 60 meters wide; this segment has a total of 14 lanes, half in one direction and half in the other, which were divided with a hazardously narrow median where pedestrians had to stop when

crossing the road. On the two opposite sides of the avenue, along the respective boulevards, two more streets, of four lanes each, complete the layout. The general state of the two boulevards and their paved spaces was quite neglected and the central median had proved inadequate for years.

The project, which was planned in four different stages, two of them fully completed so far, pursues three main objectives: the transformation of the purely car-oriented design of this urban axis, the offer of a more pedestrian-friendly environment and the revalorization of the historic planted native trees. The first stage proposed the renovation of the two boulevards as linear plazas and the enlargement of the central median.

The new plazas were largely repaved with concrete surfaces and granite pieces for their extremities, providing visual continuity between one plaza and the next. The central paved areas were completed with new urban furniture and lamp posts and the edges were replanted with herbaceous species and low maintenance shrubs, generating a carpet-like area extending underneath the large crowns of the existing trees. Whenever these were missing, as a result of the natural effects of time, the designers replanted the exact same species.

The design of every one of these boulevard-plazas responded to the pre-existing conditions of each of them, some including old fences, planters, access to the Metro or parking areas. The car accesses to public underground parking, formerly confusing and unappealing, are now marked with U-shaped green esplanades and the waiting space for pedestrians remains a top priority.

For the new median, formerly 1.2 meters-wide and now 6 meters at some portions and 8 m at others, the designers created a geometrical layout with shrubs of different heights and narrow paths that diagonally cross the planting. These paths serve as maintenance accesses and, at the same time, produce a dynamic new sense of space and a highly attractive visual effect.

Pedestrianisation

As part of the Priority for Pedestrians program, the City Government decided to start to transform some central streets, or parts of them, into pedestrian connections and

areas. This was not an easy task and, in the beginning, some of these projects were quite controversial and found important opposition groups (mostly drivers); reasons for which they are being implemented in phases and under a sensitive process of street selection. However, people's response to the new pedestrian streets has been widely positive and the program is coming together.

One of the first streets to be pedestrianized was a very short one, two blocks or 200 meters long which, although located in downtown, was originally outlined as a closed street, or passage, with very low traffic flow. The project for this street, called Tres Sargentos, sought to create one

single plane for pedestrians with only emergency and residential parking access. The original cobblestone that covered the street was left as a central ribbon and new concrete and granite surfaces were added at the sides. Bollards, linear storm drains, street trees and benches complete the new design of this now charming streetscape, ironically so close to some of the busiest avenues and streets in Buenos Aires city.

Perpendicularly connected to this street, Reconquista is another of the ones chosen to become pedestrian. Much longer and consequently more complex to implement, this one has been planned in three different phases, of four

blocks or approximately 400 meters each, the first of which has been recently finished.

Probably the most relevant characteristic of this street was the regular presence of bars and restaurants along its full extent, which has been exploited for the creation of a new inviting sense of place. The sidewalks, after being leveled with the former street, were widened, marked with bollards, furnished and planted with flowering trees.

Now, the outdoor areas of those bars integrate with the new lively pedestrian axis; office workers from this business district usually sit around here and rest during lunch and after-office hours, mingling with tourists who walk the city at all times of the day.

Transformation of Bullrich Avenue

This urban piece of approximately 800 meters long, situated in an area of the city with a high percentage of parks and landscaped avenues, used to be a bus parking and drop off area adjacent to an embanked railroad and connected with many important busy streets.

The deteriorated state of the drop off area and the general lack of maintenance of this piece turned into the main reasons to propose this piece for complete renovation, as part of the city renovation program. With the objective of transforming the place into a green oasis where people can walk and sit around, as well as a connecting urban pedestrian piece between some central highly car-oriented areas, the design focused on a simple but attractive landscape and streetscape layout.

The elevation change between the elevated railroad and the sidewalk level was approximately 7 meters and it was resolved with a steep rocky slope where clusters of plants grew spontaneously; adjacent to this, extended a dull paved surface which served as a bus parking area. As part of the original tree plantation of the city, a row of large native

1　Bullrich 大街鸟瞰图
2~4　利用天然的斜坡创造出和谐的韵律

trees, aligned in front of this embankment and facing another matching row across Bullrich Avenue, provided an important closure and a green vertical presence to the site.

Based on this existing situation, the designers decided to generate a spatial balance between original and new components of the site, thus emphasizing the longitudinal silhouette of the piece and creating a more horizontal plane, out of the existing embankment. In order to do the first, they created a series of linear planting compositions that combine different species: groundcovers, short colorful bulbs and lawn extensions when closer to the sidewalk, and taller ornamental grasses when closer to the fence that edges the railroad. These vegetal ribbons are only interrupted by a series of short gabions, perpendicularly located every 42 meters, which define a continuous rhythm and also accompany the slight differences in height of the plant compositions and the topsoil layer that each one of them requires.

These ribbons cover the area of the former embankment, whose percentage slope was smoothened, reducing it from 60% to 20%, and consequently extended toward the contiguous former drop off area. The rest of the paved surface that was occupied in the past by the drop off was adjoined to and leveled with the existing sidewalk, thus generating a much larger pedestrian space framed on one side by the old trees and on the other by the new green soft rise.

This pedestrian area was completed with new lamp posts and furnished with some fancy concrete chairs whose original design dated from the 1970s and was recreated for this and other public spaces within the central areas of the city.

The renovation process and the formal results of this centrally located urban piece were quite successful and explored and proved the need of green pedestrian spaces for people in the city.

3

卡拉沃沃大道改造

Transformation of the Carabobo Avenue

撰文：Jimena Martignoni   图片提供：EDU Medellín   翻译：高明

麦德林是哥伦比亚第二大城市，拥有 225 万人口和 38 073 万平方米的占地面积。自世纪之交以来，通过一些颇有争议的城市计划的实施与实验性建筑的建成，这个城市已经得到了彻底的改造，将以前受损的城市形象重新塑造成为代表社会进步与希望的形象。在当地政府最近连续三次选举期间提出的计划中，卡拉沃沃大道改造工程以其对人行道及林阴道的完美设计脱颖而出。

总规划图

卡拉沃沃大道是麦德林市原始格局的一部分，它沿着麦德林河向南北延伸，沿途坐落着许多20世纪10年代设计的公共设施和文化建筑。二十世纪七八十年代，随着第一批"城市试点规划"（规划一处新的商业中心，并将所有的学校机构迁移到其他区域）的部分实施，这条大道失去了其原有的重要价值而逐渐衰落。然而，基于其特殊的历史和城市意义，市政府决定对大道进行重新评估与整修，包括将大道的一部分改建为专门的人行道。这一段人行道将新商业区、整修后的城市中心区与位于麦德林北部之前被弃置如今又被翻新修复的娱乐文化区联结起来。

最早推动卡拉沃沃大道改造项目的工程之一便是对历史建筑的修缮与完整恢复。在 2000 年，曾经的市政府大楼被翻修成新的安蒂奥基亚博物馆，其邻近地区被设计成一个公共广场，用来展出国际著名艺术家费尔南多·博特罗捐赠的大量雕塑。在这些大受麦德林市民好评的工程实施之后，政府决定设计一套由三段工程构成的大道改造总方案——卡拉沃沃城市漫步改造计划。

设计师首先深入调查分析了这一地区的车流、公共交通路线和频率情况，发现公交路线最容易出现交通堵塞现象。但其中一些线路的交通情况已经得到改观，私人汽车的数量也在减少，一些车道被更改，装卸服务区也被重新划分。在大道的停车专用区，装卸工作要按照特定的时间表执行；在人行道区域，要求装载工作仅限于在辅路上进行。

根据不同的环境条件，设计将大道划分为三段，从南到北每段长度不一：第一段长 887m，第二段长1300m，第三段长 1350m。

最南端这段也是最短的一段，被称为步行街，属于被修复中心区的一部分，直接与商业区和新的光明公园相连。这个公园的标志性建筑于 2004 年建成，当时这个大道改造计划还处于初期阶段。公园一侧是新的现代公共图书馆，另一侧是两座重修的古老砖质大楼，这两栋大楼就是卡拉沃沃大道第一段南端的终点。

将这段全长 887m 的路段改造成一个商业步行街，并将其沿线的诸多商店重新组织并加以修缮翻新，是

城市改造的一项重要举措。这一决定是为了改变这一地区严重污染的环境和交通不畅的状况，为人们创造一份新的生活体验，加快地区商业一体化的进程。

从规划初期到最终落成，整个过程得到了店主们和工人们的积极配合。他们动员一切可以团结的力量，联合起来组成了社团联盟。在第一路段完成后，市政府与这些人签订了一份关于路段责任与维护的协议。

设计提出在 5.5m 宽的中央大道两旁设立小道，并且用柱子进行分隔，作为路旁商店的入口和商品展区，街道旁的新椅子和绿化使设计得到了完善。与第二段相连的最后一部分路段面向雕塑广场和安蒂奥基亚博物馆，道路一侧有块很大的空地，优美的草地点缀其间，同时作为周末集市的场所。大道上有三条机动车道，被新种植的树带隔开，大道西侧是一条人行道和一条新的自行车道。第三段也是最长的路段，随着城市进程化和社会改革的推进，这一地区已经成为了一个重要的文化和娱乐场所：麦德林植物园——这儿有新的兰花园和其他文化大楼；翻修的天文馆和与之毗邻的希望公园——这里经常举办各种文化活动和每周一次的户外电影；探索馆——这儿有一系列互动博物馆和风景优美的公共场所；还有安蒂奥基亚大学校园和北部公园。

在这片天地之中，卡拉沃沃大道为行人提供了一系列与周围建筑融于一体的步行区，使整个地区极富美感。希望公园旁边的大道由三条机动车道和一条自行车道构成；公园后面，沿着地势略高的地铁站而下，三条机动车道连成一体；公园西侧，人行道连着探索馆的草坪；公园东侧，另一条呈不规则形状的人行道连着植物园的新入口。在道路和街道表面的一些人工绘画也使道路别具一格，成为独特的景点。

麦德林这条具有重大意义的南北坐标线连接了很多近年被翻修的旧城区，这一南北轴线的改造计划已经成为城市整体形象转变的一个成功典范。

1	第二段大道可作为人行道和自行车道
2、3	被植物和桌椅装饰的新建道路

Medellín is the second largest city in Colombia, with a population of 2.25 million and an area of 147 square miles. Since the turn of the millennium, this city has been reinventing itself through controversial urban plans and experimental architecture whose main objective was the transformation of a former stigmatized image into one of social change and hope. Among these plans, proposed primarily by the local government through the last three consecutive electoral periods, the project for Carabobo Avenue rises as a very complete example of renovated pedestrian areas and promenades.

Carabobo Avenue was part of the original layout of the city of Medellín, running north-south parallel to the River Medellín and framed by many institutional and cultural buildings designed during the first decades of the 20th century. During the 1970s and 1980s, and as a consequence of the partial implementation of a first City Pilot Plan (1948) that planned a new business center and relocated all mayor institutional buildings to a different area, this avenue lost relevance and became visually and physically degraded. However, based on its particularly historical and urban significance, the city government decided to carry out a large process of revalorization and renovation, which included the transformation of part of it into an exclusively pedestrian segment. This pedestrian segment connects the new business district and the renovated central area of the city with the North District of Medellín, another formerly abandoned area and now a largely renovated recreational-cultural quarter.

One of the first actions that were implemented and acted as a trigger for the transformation of Carabobo Avenue, was the renovation and/or restoration of many of those historical buildings lying along its full extent. What had been the City Hall in the past was renovated, in 2000, as the new Museum of Antioquia (the name of the region) and the adjacent area was laid out as a public plaza in which to exhibit a large number of sculptures donated by internationally renowned artist Fernando Botero. Following these actions, which were very well received by the people of Medellín, the government decided to lay out a Master Plan for the avenue, differentiating three different segments that, together, make the Paseo Urbano Carabobo.

An initial in-depth urban study analyzed traffic flow and public transport routes and frequencies within the area and determined that bus routes would be the most

1 第三段路的俯瞰图
2 由翻新的历史建筑构成的人行道

affected. Yet, some of them had already been modified and private cars were a minor presence. As a result, some of those routes and some street directions were modified and loading and unloading service areas were newly laid out. In the segments of the avenue where vehicular lanes were conserved, loading was planned following special timetables; along the pedestrian segment's extension, loading was restricted to lateral streets.

The three segments, into which the avenue itself was divided, responded to some general existing conditions and each one covers a different length. From south to north: a first one of 887 meters, a second of 1,300 meters and a last one of 1,350 meters.

The southern and shorter segment, called Paseo Peatonal or Pedestrian Mall, is the one that connects directly with the business district and, as part of the renovated central

area, with the new Park of Light. This park has a highly institutional image, composed of 300 concrete and steel masts regularly distributed onto a largely paved plane; completed in 2004, when the plan for the avenue was in its early stages, this space is framed on one side by a new modern public library and, on the opposite, by two restored historic brick buildings. The southern extremity of Carabobo's first segment is limited by these two buildings, another reason for pursuing the integration of all these emblematic constructions and the surrounding public spaces.

The transformation of this 887 meter-long segment into a commercial pedestrian street and the reorganization and improvement of the many stores that are located along its length was a key action for this part of the city. The decision intended to change the highly polluted and inconsistent car-oriented image of the area and to create a new sense of place for people, integrating the commercial uses historically set along this street.

The whole process, from the earliest stages of planning until the last stages of detail construction, was closely followed by the different store owners and workers. They got together and created a community association through which they fostered participation of all involved actors; after the completion of the first segment, the City Government and these people signed an agreement for further responsibilities and maintenance actions.

The new design offers a 5.5 meter-wide central strip and two sides, marked with bollards, which provide access and display areas for the adjoining stores, and it is completed with chairs and newly planted street trees. The last portion, directly connected with the second segment of the avenue, opens onto the aforementioned plaza of sculptures and Museum of Antioquia, providing a much larger open area which is dotted with landscaped esplanades and provides room for the weekend fairs. The second segment runs along an area of the city which is served by many private and public buildings and which is the object of more urban renovation processes slated to be implemented over the next few years. Here, the avenue has three vehicular lanes, framed by new planted trees and, along its west side, is served by both a pedestrian border furnished with chairs and a new bicycle lane. The third and longest segment develops along the North District and enters one of the formerly most stigmatized areas of this city. Object of a profound process of urban and social transformation since the beginning of the millennium, this area has become an important cultural and recreational space: the renovated Botanical Gardens, with the new Orchid Garden and other cultural buildings; the renovated Planetarium and the adjacent Park of Wishes, where different cultural events and a weekly program of free open-air movies take place; Explora Park, with a series of interactive museums and new landscaped public spaces; the campus of the Antioquia University and the North Park.

When going through or parallel to some of these spaces, Carabobo Avenue defines a series of public pedestrian spaces that integrate with buildings and help to generate a very attractive image for the entire area. When adjacent to the Park of Wishes, the avenue offers the three vehicular lanes and a bike lane; right after this park, and passing below an elevated metro station, the avenue is widened and transformed into a composition that conserves the three lanes but adds, on its west side, a pedestrian sidewalk directly connected with the terraces of Explora Park and, on the east side, another irregularly shaped pedestrian strip that connects with the new entrance of the Botanical Gardens. Designed as part of the paved sidewalks and street surfaces, some artful patterns and drawings help to define an image that gets away from that of the typical street and generates an unusually inviting one.

The plan for the transformation of the most significant north-south axis of the city of Medellín, which connects the historically most relevant and the recently renovated areas and districts, has become a successful example of holistic urban conversion. What's more, it establishes how vital is the role of pedestrian areas within these processes and how high becomes the value added they eventually represent for any contemporary city.

自豪者之路

Proudfoots Lane

撰文：McGregor Coxall　　翻译：武秀伟

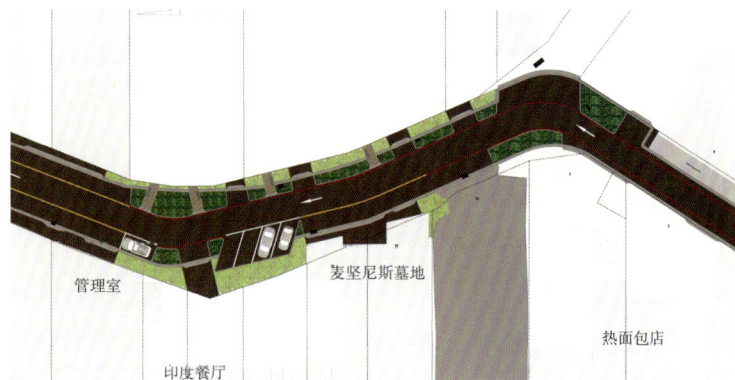

管理室

麦坚尼斯墓地

印度餐厅

热面包店

图例

- 保留下来的树
- 拟种植重新分布的吸水性强的树种
- 拟在路边种植重新分布的吸水性强的树种
- 拟种在渗透区的树种
- 拟建的公共入口标牌
- 场地的大体边界

混凝土边石	翻新后的盥洗区
标色水线（禁止停车）	系船柱
标色水线（装卸区）	长椅
标色水线（停车弯道）	木质顶棚
渗透区—穿越路面	小路（拟建）终点的延伸区域，作为周末集市的场所
拟建绿色空间	该区域可长期作为户外活动空间，也可以搭建临街画廊或工作室

总规划图（图片提供：McGregor Coxall）

通往渗透区的一角（图片设计：McGregor Coxall）

1 路边建筑（图片提供：John Mills）
2 雨水花园（图片提供：John Mills）

    Mcgregor+Partners 与特威德郡议会一起对默威伦巴中心商业区的自豪者之路进行了重新规划与设计。该方案将一处老化的后勤服务通道（主要由停车场和装载区构成）改造成一处独具匠心、风景宜人的步行区。

    新规划的目标就是鼓励当地的企业、艺术家、手工艺者和商人利用邻近自豪者之路的"奇特"空间作为工作室和艺术品陈列室，以此来恢复该通道的活力，振兴当地经济。该镇有一个十分活跃的艺术团体，而该通道有可能成为周末集市的场所。该设计利用行人踩出的近路，有针对性地改造出安全的交通道路。

    整体规划设计对通道周围的停车场和装载区进行了重组，为创建咖啡厅、工作室和社区设施提供了机会。整体规划也把以水为主体的城市设计融入其中，新设计的雨水花园就是为了过滤雨水，使雨水在流入特威德河之前，就能将污染物过滤掉。改良后的雨水花园有助于街道上的树木进行自我灌溉（它们可以凭借路边的雨水进行灌溉）。每一处雨水花园都有一个平台和长椅供人们休憩、午餐或等人。该通道主要通行卡车和服务性车辆，因此，雨水花园可以有效地阻止大量油污流入河中。

    该项目规模小、预算低。但它却证实了当工程师们支持采用非传统的设计方法时，他们将会受益匪浅。

Mcgregor Coxall worked with Tweed Shire Council to develop a strategy and master plan for the redevelopment of Proudfoots Lane in the centre of the Murwillumbah CBD. The brief prepared by the Council's landscape architect entailed taking a degraded back service laneway dominated by carparking and loading zones and regenerating it into a pedestrian friendly artisan precinct.

The strategy aimed to rejuvenate the laneway and local economy by encouraging local businesses, artists, craftsmen and tradesmen to use the many "quirky" spaces fronting the laneway as workshops and galleries. There is a vibrant arts community in this country town and the laneway has potential for weekend market use. The design capitalizes on the main short cuts used by pedestrians and makes safe pathways for movement.

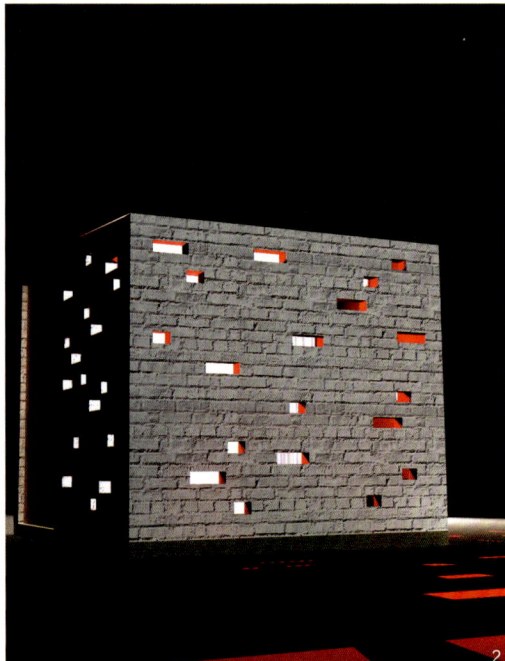

1 原有的墙（图片提供：John Mills）
2 用灯装饰的建筑（图片提供：McGregor Coxall）
3 雨水花园中升起的长椅（图片提供：John Mills）
4、6 雨水花园（图片提供：John Mills）
5 坐椅（图片提供：John Mills）

The master plan design re-organises the parking and loading zone requirements of the laneway to create opportunities for cafes, workshops and community facilities to be located in the precinct. The master plan also incorporates water sensitive urban design with new rain gardens for storm water infiltration designed to filter pollutants from the road before they enter the nearby Tweed River. The retrofitted rain gardens support street trees that are self-watered by the road run off and each garden has an accompanying small deck and a seat encouraging people to stop and relax, eat lunch or simply wait. As the lane is used primarily by trucks and service vehicles, large amounts of oil have been prevented from entering the river through use of the simple rain gardens.

The project is small in stature and low budget but it demonstrates what can be achieved when council engineers are willing to support non-traditional design methodologies.

街头艺术的功能化 —— 卡克斯顿罗马行人通道

Making Street Art Functional — Caxton Roma Pedestrian Link

撰文：Pedro F Marcelino　　　图片提供：EDAW　　　翻译：刘丹春

1　布里斯班河航拍图
2　布里斯班商务中心区和故事桥
3　桥边观景玻璃窗

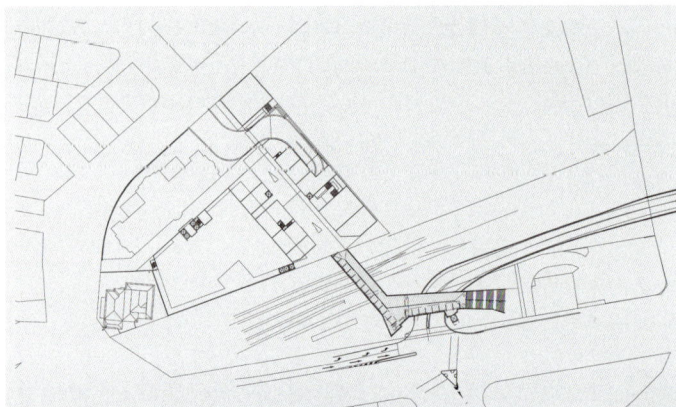

规划图

布里斯班是昆士兰州的首府，也是该州最大的城市，位于澳大利亚北部。因为人口不足两百万，它总被认为不像悉尼或者墨尔本那么国际化，而其别称布里斯维加斯暗含着能源丰富和文化繁荣之意。当地的土著居民把这个地区叫做 Mian-jin（尖峰状的地方），大概是因为它坐落于布里斯班河畔，位于莫顿湾和昆士兰州东南部大分水岭之间一个地势低洼的河漫滩之上的缘故。

卡克斯顿罗马行人通道由昆士兰州政府委托 EDAW 进行设计，将布里斯班著名的卡克斯顿街区和罗马街上区连接起来，也是一座将拥有 52 500 个坐席的桑科体育场与罗马街车站和市中心连接起来的桥。

这座桥横跨布里斯班最繁忙的铁路网络路段之一，这给施工带来了很大的挑战。为使封闭的铁路最小化，建筑元件均在建筑工地之外制造，然后进行组装并成套地安装到桥面板上。铁路也影响了建筑材料的选择和细节的设计，带有开口的钢板和如铁轨般狭长的窗户可以让大人和孩子们看到下面疾驰的列车。

站在桥中心的人们透过镶有玻璃的最高的瞭望台向外望去，可以将视野更广阔的布里斯班河、山脉、现代艺术馆以及 Four-X 啤酒厂的标志尽收眼底。设计师经过深思熟虑，决定使作品在满足功能需求的同时也要蕴含美，这就产生了令人称奇的综合艺术品。玻璃屏可以防止物体掉落到下面的铁轨上，而人物的图像被嵌在开口的钢板中，使整座桥即使在没有比赛的日子里也感觉"人很多"。

因桥与铁轨太接近而严重影响到了材料的选择和建筑细节的设计。为实现铁路架线系统的功能美，设计师选择坚固的垂直结构、简洁的连接以及含蓄的银灰两色。从桥体到罗马街上区的巨大阶梯用砖建成，以此来巩固自身的结构，并与卡克斯顿街和皮特里·苔瑞思街区现存的砖石建筑相搭配。

举行重要活动的时候，桥面板和台阶需要在很短的一段时间内承受密集的人流，这就要求桥面的最小宽度为 8m。当然，使用强度和可预见的密集人流等特点也影响了材料的选择和建筑细节。建桥不是为了满足提供阴凉的需求，但是设计团队认为遮阴很有必要，于是慷慨地建造了遮阴篷。

作为内城北部公交枢纽联盟的一员，EDAW 承接了该桥的设计。设计师在早期便与结构工程师和公交路线设计师紧密合作，确定桥面板的高度、层次和落差，这也是设计方案符合技术需求的决定性因素。

通过内城公共交通的顺畅运行，这条重要的行人通道承担着减少布里斯班私家车使用量的责任，因为经由此道，穿梭于罗马街火车站、内城北部公交车站和经过市中心的所有公交线路之间均不超过 5 分钟路程。

这项朴实的工程凭借其自身简洁、迷人和卓有成效的设计取得了巨大的成就。

Brisbane is the state capital of Queensland, in northern Australia, and its largest city. With a population of nearly 2 million people, it is often said not to be as cosmopolitan as Sydney or Melbourne, yet its nickname of Bris Vegas hints at plenty of raw energy and cultural buzz. The local indigenous people knew the area as Mian-jin ("place shaped as a spike"), probably owing to its location on the Brisbane River, on a low-lying floodplain between Moreton Bay and the Great Dividing Range in south-eastern Queensland.

Connecting Brisbane's historic Caxton Street precinct with Upper Roma Street, Caxton Roma Pedestrian Link was commissioned to EDAW by the Queensland State Government as a bridge linking the 52,500-seat Suncorp Stadium to Roma Street Station and the inner city.

The bridge crosses one of the busiest sections of Brisbane's rail network, posing challenges to construction. To keep track closures to a minimum, the architectural elements were designed as a "kit" of parts that could be manufactured off-site and then assembled and installed on the bridge deck itself. The railway also influenced the choice of materials and design details; perforated steel screens and long "train track" windows allow both adults and children to view the rolling stock travelling below.

Wider views of the Brisbane River, the mountains, the Gallery of Modern Art and the iconic Four-X Brewery, can all be experienced from the full-height glazed lookout in the centre of the bridge. A deliberate decision to draw beauty from functional requirements has also resulted in a whimsical integrated artwork. Throw screens were required to prevent objects being tossed onto the railway tracks

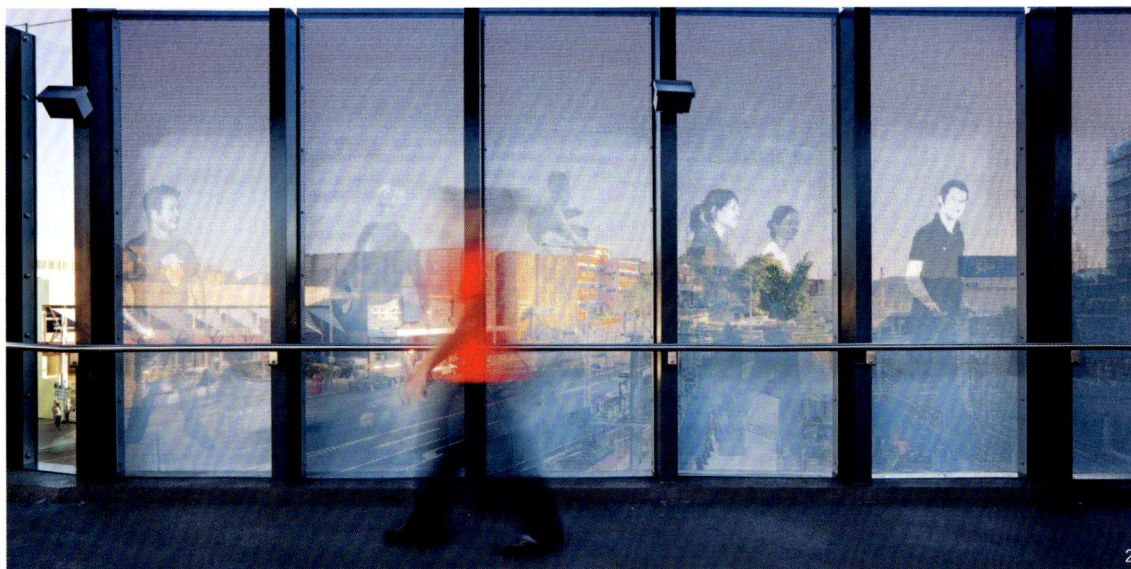

1 砖石构建的台阶
2、4 嵌在玻璃屏中的人物图像
3 桥下即是城市主要交通线

beneath, whereas Images of people were incorporated into perforated steel panels, allowing the bridge to feel "populated" even on non-match days.

The bridge's close proximity to the rail lines heavily influenced material selection and detailing. The functional beauty of the rail catenary system informed the choice of a robust vertical structure with simple expressed connections and subtle silver and grey colours. The large staircase from the bridge to Upper Roma Street was treated in brick to ground the structure and connect it with the existing brick buildings in the Caxton Street and Petrie Terrace precincts.

The bridge deck and stair accommodate a high volume of people in a short period of time following major events, prescribing the minimum width of eight metres. Naturally, the intensity of use and nature of anticipated crowds also influenced the choice of materials and detailing. It was not a brief requirement to provide shade, but the design team considered this essential and introduced a generous canopy.

EDAW undertook the bridge design as a member of the Inner Northern Busway HUB Alliance. Close and early collaboration with the structural engineers and Busway designers influenced the heights, levels and falls of the bridge deck, determining that design solutions clearly answered to technical needs.

The construction of a significant pedestrian bridge located within five minutes walking distance of Roma Street rail station, Inner Northern Busway stations and all bus routes servicing the central city demonstrates serious commitment to reducing private car use in Brisbane through the promotion of public and active transport in the inner city.

Instantly utilized by Brisbaners, this modest project achieved big results, by keeping design simple, attractive and efficient.

上海古北金街

Shanghai Gubei Gold Street

撰文：SWA Group　　图片提供：Tom Fox　　翻译：曹亮

　　古北金街位于上海这座建筑物密集的大都市的中心，是一个以行人为主的多功能露天场所。古北金街跨越4条街，长约800米、宽40米～80米，两侧是20层的高层住宅群，这些高层的底层设计为商店、餐馆和精品展示长廊。

　　该项目的设计理念是以大规模的景观元素为特点，在高层建筑群中为人们提供一个舒适的、人性化的环境。古北金街被与它相交的两条街分成三个主要部分：第一部分的底层设计成餐馆和咖啡厅；在第二部分的底层设计成面向社区的服务设施；最后，在第三部分的底层设计成高档的零售店以及艺术品和工艺制品的展示长廊。该项目的景观元素包括一条主步行街、一个露天的餐饮小岛、一个眺望平台、一个大型的社区广场，还有一些小型的可用于商品零售的庭院。广场、咖啡厅、餐饮和零售用的平台、圆形的小广场和一个由树木围成的有纪念意义的凸起平台共同成为人们关注的焦点和活动的空间。这条步行街不仅提供了公共开放空间的新体验，同时也挖掘了在密集型城市中构建生态环境的潜力。

　　步行街以成排的银杏树为特色，这些银杏树井然有序的布局与庭院和花园里不拘形式种植的观赏树丛形成对比。通过种植多种多样的植物、设计风格不同的坐椅以及铺设图案抽象的石砖路，使得每一个供行人休息的小庭院都有自己的特色。而精心选择种植在边缘的植物及细心设计植物的高矮变化，仿佛为地表铺设了一块纹理精细的地毯。设计师同时考虑到利用植物的季相变化来体现四季更替。

　　该项目现已成为市民聚会、漫步的好去处，设计成功地为上海中心区域注入了新的生机和活力。

1 古北金街位于鳞次栉比的建筑物的中心，周围为
风格迥异的景观花园
2 铺砖的图案自然流畅地连接了建筑物和露天广场
3 景观设计力图将室内外空间自然流畅地联系在一起

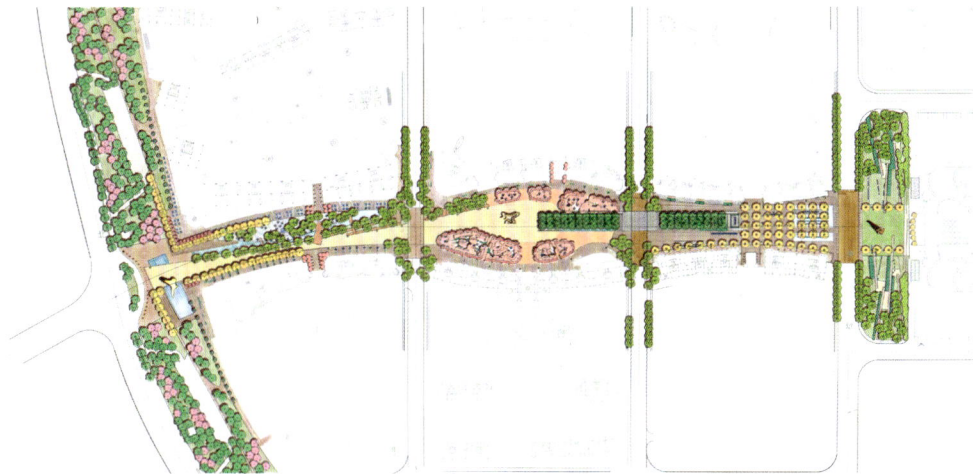

总体规划图

Shanghai Gubei Gold Street is a mixed-use pedestrian-oriented open space in the midst of the densely built city. The site spans four city blocks, is approximately 800 meters long and 40-80 meters wide, and is flanked by 20-story residential towers with shops, restaurants, and galleries at the ground floor.

The SWA design concept for Gubei Gold Street is defined by large-scale landscape gestures that provide a comfortable human scale environment amidst highrise towers. The site is comprised of three major parcels, and is intersected by two neighborhood streets that divide the site into three main sections. The SWA design for the site consists of restaurants and cafes on the first block, community-oriented facilities on the second block, high-end retail and arts and crafts galleries on the third block on the ground level. Landscape elements include a major pedestrian promenade, an outdoor dining island, a viewing terrace, a grand community plaza, and smaller retail courtyards. Plazas, cafe, dining and retail terraces, an amphitheater, and a monumental tree-lined raised terrace attract focus and activity. The pedestrian mall offers a unique public open space experience as well as potential for urban ecology in a dense urban setting.

The promenade is defined by tree rows of Gingko, a more formal arrangement that contrasts with informal groves of ornamental trees planted in the courtyards and parks. Each of the small court seating areas has a unique character. This is achieved through the varieties of plant material, custom designed seating and abstract patterned stone paving. The planting design creates a subtle carpet of texture at the ground level, with great attention to the selection of plant materials in the border structure through the careful selection of varying plant heights. There is much attention given to the seasonal changes in the plant color, ornamental trees mark the changing season in foliage and leaf color.

The Shanghai Gubei Gold Street Project with its many parks and promenades has become a popular gathering place for city residents at all times of the day and into the evening. The success of the project has brought new energy to this part of the city of Shanghai.

1　步行街为市民提供了公共开放空间

2　蕴含民俗元素的设施体现出了人性化的设计，
而银杏树柔和了周围高层建筑与广场之间的过渡

3　广场成为市民休闲娱乐的好去处

4、5　银杏树遮蔽的步行街是人们钟爱的漫步地点

6　步行街成为了标志性的景观区

7　小型庭院夜景

Nicholson 步行街

Nicholson Street Mall

撰文：HASSELL　　　图片提供：Dianna Snape　　　翻译：董桂宏

该项目地处墨尔本市 Footscray 的中心，这里是墨尔本市的文化中心和主要商业区，代表着 Footscray 的文化和商业发展水平，也是 Footscray 未来经济发展的核心区域。

为满足当地多种多样文化活动的需要，这一区域的发展规划着重强调步行街的综合服务功能。步行街建成于 20 世纪 70 年代，是澳大利亚首个商业步行街。虽然建设商业步行街在当时是一项创举，但是随着经济的发展，这个步行街已渐渐不能满足商业和文化发展的需要。2006 年，墨尔本市议会决定重新开发这一地区。HASSELL 公司对步行街重新进行了规划，并参考国际上最新的城市公共空间规划案例，制定出能够振兴该城市商业区的设计方案。场地上原有的高大树木具有遮阳功能，其中间隔种植着落叶树，以增加场地冬季的光照面积。同时设计师还引进了许多本地的耐旱植物用以美化环境。场地的坐椅和其他设施尽量

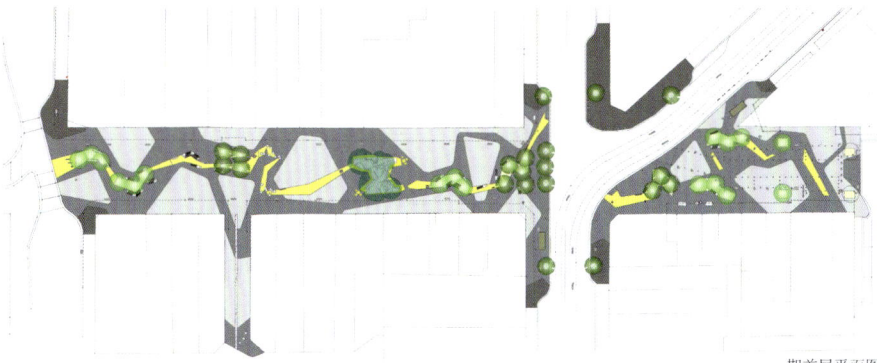

一期首层平面图

都设置在树阴处，为人们休憩交流提供良好的环境。在全新的规划方案中，Nicholson 街北部首次被纳入步行街的范围内，大大增强了步行街的文化功能，到处洋溢着都市的休闲文化气息。

该项目拥有先进的照明设备和富有情趣的坐椅，成为人们休闲聚会的理想场所。人们三五成群地来到这里，或逛街、或聚餐、或倾心交谈，轻松自在。步行街的地面上"飘荡"着一条黄色的"丝带"，贯穿步行街始终，既美化了步行街的景观，又能够对步行街的功能性元素进行分类（如街道家具和装饰等），具有引导的功能。形象的黄色"丝带"是一项创新的设计，其亮丽的色彩和动感的线条充分反映出当地的特色文化活动——生机勃勃的街头文化、具有独特魅力的街头艺术、频繁的庆典活动和舞会等。根据顾客的需要，定制的混凝土铺装选用了优雅的浅灰色和深灰色，而定制的混凝土坐椅的折角与黄色"丝带"的折角大小相同，相互呼应。该项目不仅是生意繁荣的商业区，也是人们举行聚会和文化活动的理想场所。

设计的最大挑战是如何使步行街体现 Footscray 独特多样的文化特征。在多方的共同努力下，该项目成功地克服各种了挑战，并最终发展成为 Footscray 重要的商业和文化中心。该设计最鲜明的特色是简洁的黄色"丝带"设计，这条"丝带"不仅将步行街的各个元素串联起来，形成和谐的整体；并仿佛是项目的驱动程序一般，从设计到施工一直引导着步行街的重建工程。同时，这个项目也堪称是当代商业中心改造项目中的典范。如今，Nicholson 步行街的商业繁荣、文化兴盛，其中也饱含着改造工程的一份力量。

Nicholson Street Mall is located in the centre of Footscray, Melbourne. It is the primary civic space for the local community serving as its commercial and social hub and representing the identity and aspirations of both Footscray and the entire municipality.

The space needed to facilitate a strong sense of community and cater for a diverse range of cultural activities. The previous condition of the mall did not reflect its importance and in 2006, Council committed to revamping this important place. Built in the 1970s, the pedestrian mall was the first of its kind in Australia and represented an innovative approach in its creation of a car-free civic space. The HASSELL redesign continues this legacy of innovation through the installation of a vibrant and robust civic space that incorporates the latest initiatives in water sensitive urban design and public realm design. Existing healthy large trees ensure continuous shade in the mall, and deciduous trees have been incorporated to maximise winter solar access. Drought tolerant natives have also been introduced. Seating and other street furniture have been co-located with trees wherever possible. The northern part of the project area has been added to the pedestrian mall and therefore enhanced the character of the space as a meeting

1、2 Nicholson 步行街夜景 2
  3 定制的混凝土坐椅
  4 步行街上巧妙摆放的各种设施
  5 Nicholson 步行街街景 2
  6 人们在此休憩闲谈

place for the community.

The mall now has ample lighting, open views and attractive seating. It is an important gathering space for a range of community members, providing a place to congregate, to converse and listen, to perform and watch, to eat, drink and go shopping. A yellow ribbon, representing a "line of intensity", runs through the mall and is used as an organisational device to group all the functional elements such as street furniture and amenities. The custom-made concrete seats reflect the angular forms appearing in the yellow ribbon. The dynamic, graphic and bold design references the colour and movement of local cultural practices—vibrant and active street life, dance, celebration and street art. Light and dark grey pre-cast concrete pavers were used to differentiate between spaces and enhance the graphic language of the design. The design incorporates intimate seating clusters while accommodating community activities such as cultural events, markets and gatherings.

It was a challenge to respond to the diversity and uniqueness of Footscray. Nicholson Street Mall represents a successful collaboration between the designers, contractors and fabricators, council, engineers, artists and the local community to achieve a unique and sensitive reworking of Footscray's central civic space. There is a clear connection between the preliminary design work and the built outcome. The conceptual structuring element—the abstract yellow line—has been a clear driver in the resolution of the design through the design development phase and is clearly evident as the unifying element to the project on site. It is a clear example of how landscape architecture can provide direct benefits to the local community through the enhancement of civic spaces at a local level. The project represents a contemporary and provoking update of the original mall design and the result is an outstanding built outcome.